The Mystery of the Moon Illusion

Ptolemy of Alexandria surveying the celestial bodies, accompanied by Astronomy. (Gregor Reisch, Margarita Philosophica, Basel, 1508. Nr 45, Sektion C: Humanisten und Neulateiner, Katalog VIII, Tafel 6. Courtesy of the Warburg Institute, University of London.)

The Mystery of the Moon Illusion

Exploring Size Perception

HELEN E. ROSS
University of Stirling

CORNELIS PLUG,
University of South Africa

OXFORD
UNIVERSITY PRESS

Great Clarendon Street, Oxford OX2 6DP

Oxford University Press is a department of the University of Oxford.
It furthers the University's objective of excellence in research, scholarship,
and education by publishing worldwide in

Oxford   New York
Auckland Bangkok Buenos Aires Cape Town Chennai
Dar es Salaam Delhi Hong Kong Istanbul Karachi Kolkata
Kuala Lumpur Madrid Melbourne Mexico City Mumbai Nairobi
Sao Paulo Shanghai Taipei Tokyo Toronto

and an associated company in Berlin

A catalogue record for this title is available from the British Library

Library of Congress Cataloging in Publication Data
(Data available)

ISBN 0 19 850862 X

10 9 8 7 6 5 4 3 2 1

Typeset by Cepha Imaging Pvt Ltd, India

Printed in Great Britain

on acid-free paper by
Biddles Ltd, Guildford & King's Lynn

Foreword

This scholarly book explores a famous illusion of visual space that has been discussed by philosophers and scientists through millennia of time. Why does the moon look larger low on the horizon than high in the sky? Is this physics, physiology, psychology – or what? There have been perhaps a hundred explanations and no one is quite certain even now – though it does tell us more of mind than matter.

This is the record of a journey with expert commentary. The experts on perception will find their pet ideas, at least in embryo form, in the antique sources; philosophers will be teased by conflicts of appearance and reality; astronomers will find that they are part of the universe after all. It is the details that make this book fascinating, as they paint a picture of how we see – and get the heavens wrong, with our eyes trained for earthly things.

Richard L. Gregory

Credits

Credits for figures are given in the captions.

Credits for literary extracts are as follows:

p. 142. The translation of Castelli's (1639) work by P. E. Ariotti (1973b) is reprinted by permission of the *Annals of Science* (http:://www.tandf.co.uk).

pp. 195–196. Extract from HUMAN CROQUET by Kate Atkinson published by Black Swan (original imprint Doubleday). Used by permission of Transworld Publishers, a division of The Random House Group Limited.

p. 198. The poem ACADEMIC MOON by Helen Bevington is reprinted by permission of her estate (copyright Helen Bevington 1946) and Houghton Mifflin. The poem was originally published in *The New Yorker* and was reprinted in Bevington, H. S. (1950) *Nineteen million elephants, and other poems*. Boston: Houghton Mifflin.

Preface

This book has taken many years to write. The main reason for its slow progress was that we wished to study the history of the moon illusion in some detail, and historical research is usually slow. The tendency among contemporary investigators to ignore observations and theories from the past 24 centuries has given rise to many misconceptions, and to frequent reinventions of the same (usually inadequate) explanations. We hope that the present book will help to give due recognition to earlier work on the illusion and put it in its proper perspective.

One of the benefits of the delay in completing the book is that we were able to consult some important new sources. These included English translations of Ptolemy's *Optics* (Smith 1996) and of the *Optics* of Ibn al-Haytham (Sabra 1989), as well as recent developments in perceptual theory that underlie possible explanations of the moon illusion.

We started the book with the aim of reviewing all the main explanations of the moon illusion in a historical manner, and without presupposing any particular solution. However, some accounts had already been disproved by earlier authors, while others could be rejected with reasonable certainty on the basis of current knowledge. It became clear that several factors contributed in various degrees to the illusion. We therefore decided that the best solution was to quantify the relative contribution of the different components – the most important of which is the sight of the intervening terrain.

The history of the moon illusion to a large extent reflects the history of theories of size perception. One of the difficulties in writing this book was to decide how far to include relevant material on size perception that was not strictly about the moon illusion. We have included some such material, but if we strayed further the book would be twice the length and half as readable. We hope we have indicated enough references, so that interested readers can follow up these sidelines. To avoid too much clutter in the main text we have relegated secondary references and more abstruse material to endnotes; but all references are given in full at the end of the book.

We should like to thank Michiako Kondo for translations of Japanese and Chinese material; Loek Schönbeck and George MacDonald Ross for some classical and philosophical references; Nicholas Wade for unearthing Thomas Young's diagram of the sky illusion; Peter Hancock, Peter Hucker, and Lloyd Kaufman for assistance with illustrations; and Bill Barnes-Gutteridge, Alan Benson, and Robert O'Shea for comments on individual chapters. Our special thanks go to Lloyd Kaufman, who commented on much of the book and suggested many improvements. Helen Ross would also like to thank Richard Gregory for inspiring her interest in celestial matters;

the University of Stirling for sabbatical leave; and the Leverhulme Foundation for a travel grant to work with Cornelis Plug.

Stirling and Pretoria 2001

Helen E. Ross   Cornelis Plug
Department of Psychology   Department of Psychology
University of Stirling   University of South Africa
Stirling FK9 4LA   PO Box 392 Pretoria 0001
Scotland   South Africa
h.e.ross@stir.ac.uk   plugc@mweb.co.za

Contents

CHAPTER 1

The celestial illusions

When the full moon rises above the horizon it usually appears much larger than when it is higher in the sky. The same phenomenon occurs for a sickle moon, but it has attracted less attention than when the moon is nearly full. The difference in size is normally experienced as an enlargement of the horizon moon, rather than a diminution of the raised moon. It is easy to convince oneself that the size of the moon does not change appreciably with height, by comparing its angular size to that of the tip of one's little finger held at arm's length. This informal experiment will convince most people that the enlargement is apparent rather than real, but the conviction will not cause the illusion to disappear.

The apparent horizon enlargement of the moon is usually called the *moon illusion*, a term which became popular in the twentieth century.[1] It has also sometimes been called the *horizon illusion*,[2] but this is clearly less appropriate. A similar illusion can be observed for the sun, and it is normally called the *sun illusion*.[3] This illusion is less well known to most people than the moon illusion, perhaps because the sun is usually too bright to observe with the naked eye. However, the sun illusion was more frequently discussed than the moon illusion in the early literature. Although two different celestial bodies are involved, there seems to be no fundamental difference between the two illusions: they are therefore generally considered to be two examples of the same phenomenon. A third example is the apparent enlargement of the constellations and of the distances between the stars near the horizon. In the words of the astronomer Paul Stroobant (1884, p. 719): 'The same phenomenon exists for the constellations; thus the Great Bear and Orion, close to the horizon, appear enormous.' This form of the illusion is probably the least observed of the three, since fewer people give serious attention to the night sky. The term *celestial illusion* has been proposed to cover all three examples.[4]

Most people are aware of the celestial illusion in some form or other through personal experience. Enright (1975, p. 87) comments that the moon illusion 'is probably the best known and most frequently discussed of all optical illusions, since it occurs in a natural setting, and represents a compelling, large-magnitude distortion of reality'. Children experience the illusion as well as adults,[5] and the phenomenon has been noted in many cultures. It was experienced by some of the islanders in the Torres Straits, according to the Cambridge Anthropological Expedition at the end of the nineteenth century,[6] and it has also been reported for many different times and places. The experience is so universal that artists often enlarge the horizon moon in their paintings, a topic that we discuss in more detail in Chapter 4.

There have been reports of other illusions in the sky which may be related to the celestial illusion. One of these is the apparently flattened form of the sky dome (Chapter 8). Another effect concerns the shape of haloes around the sun or moon. Such halos are usually circular, with the sun or moon at their centre; but they sometimes appear egg-shaped, with the longest axis perpendicular to the horizon and the blunter point below. In addition the sun may appear to be above the centre of the halo. This appearance is what one would expect if the lower part of the circular halo were enlarged relative to the upper part; and in fact it has been reported that the whole halo appears to shrink as it rises higher above the horizon.[7] One should, however, be careful of interpreting all the non-circular appearances of haloes as illusions. At least some haloes are measurably elliptical.[8] A similar distortion is sometimes observed in the rainbow, which may appear to be broader closer to the horizon than higher up.[9]

Historical perspective

In this book we often refer to observations made a considerable time ago, for several reasons. In the first place, the scientific study of the illusion has been in progress for a long time, so that many early observations are available. Some were careful and extensive studies, and their results remain useful. Others were more casual observations, but they help us to appreciate the variable size of the illusion experienced by people with different backgrounds and under different circumstances. Second, most current theories of the illusion have ancient origins, so the older literature has retained its relevance. Finally, the history of speculations and research on the celestial illusion is of interest in its own right, and also reflects many important aspects of the history of visual perception.[10]

General awareness of the illusion dates back quite some time. In 1762 the Swiss mathematician Leonard Euler began one of his scientific letters with these words: 'You must have frequently remarked, that the moon, at rising and setting, appears much larger than when she is considerably above the horizon; and every one must give testimony to this phenomenon.'[11] Similar thoughts were expressed during the same year in an address before the Royal Society of London by Samuel Dunn: 'The sun and moon, when they are in or near the horizon, appear to the naked eye of the generality of persons, so very large in comparison with their apparent magnitudes, when they are in the zenith, or somewhat elevated, that several learned men have been led to enquire into the cause of this phenomenon.'[12]

In fact the celestial illusion appears to have been well known for at least 23 centuries. The first known scientific mention of the illusion is by the Greek philosopher Aristotle (384–322 BC). In his book on celestial matters he briefly mentions the horizon enlargement of the celestial bodies, ascribing it to an effect of the atmosphere (see Chapter 5). In later classical times the illusion was considered by various authors, including the second-century astronomer Claudius Ptolemaeus (or Ptolemy).

Although the scientific study of the illusion can be said to have started with Aristotle, there may be an even earlier reference to it. It occurs in cuneiform script on one of the clay tablets from the royal library of Nineveh and Babylon. The library is associated with king Assurbanipal of Assyria (668–626 BC), but many of the tablets probably date from earlier times. The tablets contain astrological reports, which include observations of the celestial bodies and the weather. On tablet number 30 we find the following: 'When at the moon's appearance its right horn is long and its left horn is short, the king's hand will conquer a land other than this. When the moon at its appearance is very large, an eclipse will take place. When the moon at its appearance is very bright, the crops of the land will prosper.'[13] This seems to be the earliest reference known at present which may refer to the illusion.[14]

The 'appearance' of the moon in this text could refer either to its rising or to its first visibility after new moon. The first statement, describing a sickle moon, suggests the latter. However, the second statement may refer to a rising full moon that does not always appear the same size, presumably due to variations in the moon illusion. The observed variation in its size was clearly accepted as real, and just like the variations in brightness and orientation of the sickle moon, it was taken as a physical sign for predicting the future. The intellectual tradition at the time did not allow a distinction between apparent and real phenomena in the sense that we use these terms today: the horizon enlargement of the moon was accepted as an observational fact, as it was for several centuries to come.

There seems to be no direct evidence from earlier times relating to the moon illusion, but it is likely that the enlargement of the sun and moon was observed even in prehistoric times. In his recent book on astronomy in prehistoric Britain and Ireland, Ruggles writes:

> We have mentioned that the moon appears larger when it is close to the horizon. As a result it may be perceived as closer, and so we should explore how distance was perceived in other cultural contexts and ask whether special significance might have been attached to the moon 'when it was near'. Might this explain the apparent interest in Bronze Age western Scotland and south-west Ireland in the southern moon near the major standstill limit, when the moon never rises far above the southern horizon?[15]

A *major standstill* occurs every 18–19 years, when the pattern of swings of the location of moonrise and moonset is at a maximum, and is repeated over several months. At the southern extreme, the rising and setting locations of the moon are at their closest to each other. Diodorus of Sicily, a Greek historian from the first century BC, may well have been referring to the Callanish stone circle in the Outer Hebrides[16] when he described a fertile northern island, no smaller than Sicily, beyond the land of the Celts, where there is:

> a notable temple which is adorned with many votive offerings and is spherical in shape … the moon, as viewed from this island, appears to be but a little distance from the earth … the god visits the island every nineteen years, the period in which the return of the stars to the same place in the heavens is accomplished … At the time of this appearance … the god … plays on the cithara and dances continuously the night through from the vernal equinox until the rising of the Pleiades.[17]

Figure 1.1 Telephoto picture of the southern moon viewed through the prehistoric stone circle at Callanish, at the major standstill in June 1987. (Copyright Margaret Curtis. Reproduced with permission.)

A photograph of the low southern moon viewed through the Callanish stones is shown in Fig. 1.1. We do not know if the early worshippers were impressed by the size of the moon when near the stones, but modern humans certainly are. Robert McNeil (2001) recently watched the moon rise at this spot and wrote: 'Suddenly, on the horizon, we saw a thin red line of light. Then a great big crimson moon with a belt of cloud round its waist rose up with quite astonishing speed between the standing stones. It was one of the most fantastic things I had ever seen in my life.'

The literary tradition

Our main literary sources for the celestial illusion begin with the Greek authors in the fourth century BC. At that time Athens was the main academic and cultural centre of the western world. When Alexander the Great died in 321 BC, one of his generals (called Ptolemy) made Alexandria in Egypt his capital city. This Ptolemy and his successor assembled a magnificent library there, containing perhaps 500 000 parchment rolls and including the literature of many nations. Eratosthenes, the first person to make an accurate calculation of the size of the earth, was the chief librarian from 235–195 BC. Alexandria continued as an intellectual centre for several centuries, but fell into decline. The library was first partly burned when Julius Caesar took Alexandria. In AD 389 it was again partly destroyed by Christians under bishop Theophilus, and what remained was burnt by the Arabs in AD 640. Many early scientific works were lost without trace. There must have been many textbooks on astronomy and optics that mentioned the celestial illusion, but we do not know the names of their authors. The texts that have survived are not very helpful in this respect, since it was not customary to cite all earlier sources.

Despite their previous disrespect for other people's literature, it was the Arabs and the Christians who preserved the texts that we have today. The Arabs translated the works of Aristotle, Ptolemy (the second-century scientist, not the general), and other classical authors into Arabic before the tenth century, so they were familiar with classical explanations of the illusion. One of these Arab scientists, Ibn al Haytham (or Alhazen, 965–1039), developed quite original views on the relation of the illusion to the flattened appearance of the sky – views which have been actively discussed ever since. His book on optics, containing his theory of the moon illusion, played an important role in the scientific tradition of medieval Europe.[18] Its influence was largely due to three works on optics that were based on it: these were written in Latin about the same time by the English Franciscan friar Roger Bacon (*c.* 1263), the Archbishop of Canterbury John Pecham (*c.* 1274), and the Polish physician Witelo (*c.* 1270). Latin translations of many Arab works (which included earlier classical texts) became available to European scholars during the eleventh to thirteenth centuries. Copies of these translations were preserved in Christian monasteries and colleges, which acted as institutes of learning. Out of these colleges grew the modern independent universities.

The influence of classical and Arab authors on later accounts of the celestial illusion is quite unmistakable. Thus we find Ptolemy's explanation (in terms of optical refraction by the earth's atmosphere) repeated in several medieval and later works. Aristotle's brief account has also been paraphrased quite often, while the influence of Ibn al-Haytham is still present. During the Renaissance the problem was considered in several notes by the Italian artist and inventor Leonardo da Vinci (1452–1519). Interest in the subject, as in so many others, reached a peak during the seventeenth century, when it attracted the attention of several prominent figures in the history of science such as Johann Kepler (1571–1630), the German astronomer famous for his laws of planetary motion; René Descartes (1596–1650), the French philosopher and mathematician; and Christiaan Huyghens (1629–95), the Dutch physicist and astronomer.

After the scientific renaissance of the seventeenth century the number of publications dealing with the celestial illusion declined. It then rose from about the mid-nineteenth century to a more pronounced peak around 1900–10. This work was mainly of German and to a lesser extent French origin. At about this time the main method of investigation changed from observation in natural settings to experimentation, both in the laboratory and in the field. Interest in the illusion then declined but was followed by a third peak during the 1960s and 1970s, representing mainly American research.

Changing disciplines

The history of the illusion also shows how attention to the problem has gradually shifted from one discipline to another. During the earliest times scientists and philosophers had wide-ranging interests, and were not clearly divided into different disciplines. By the seventeenth century the distinctions were becoming clearer, and the renewed interest in the illusion came mainly from astronomers: this interest continued until they had proved to their satisfaction that the phenomenon was not astronomical. The second wave of interest, from about 1860 to 1920, came mainly from scientists in physiological and meteorological optics. This lasted until it became clear that the problem did not belong in these fields either. The third phase, which is still continuing today, comes mainly from psychologists. This may be the final phase, as the illusion clearly lies in the realm of visual perception. However, the explanation of visual perception is being taken over by neural scientists, so ownership of the problem may pass to them in the twenty-first century.

Historical changes in the study of the illusion follow the development of knowledge about perception. Three stages of knowledge can be distinguished: first, a belief in direct perception; second, an enquiry into how sensory stimulation produces perception, and how illusions arise; third, further enquiry into how perceptions are constructed from ambiguous or inadequate sensory information.[19]

The earliest view, held during ancient times and by some medieval authors, was that perceptual knowledge is directly impressed upon the senses by stimulus objects. The study of perception was mainly concerned with the problem of how a stimulus object produces a sensory stimulation. The retinal image was not understood, and it was assumed that reality is perceived directly and accurately by the eye. This view implies that all illusions are physical phenomena. Astronomers at this stage were concerned to show that the celestial illusion was an astronomical or optical phenomenon.

The second phase began with the distinction between appearance and reality, and an interest in the causes of perceptual illusions. The distinction between perceived size and visual angle was probably current by the first century BC, and was stated by Ptolemy (second century AD) and Cleomedes (c. early third century AD) . More detailed discussions of the nature of vision and perceptual illusions can be found in the works of Ptolemy and Ibn al-Haytham. Thus the second phase had early beginnings that overlapped the first phase. At this stage, scholars were beginning to offer physiological and psychological explanations for the celestial illusion, often alongside atmospheric explanations.

In the third phase, perceptual knowledge was regarded as a hypothetical construction from insufficient stimulus information. This phase was typified by the German physiologist Hermann von Helmholtz (1856–66), who regarded perception as an 'unconscious inference'. Psychology did not become a recognized discipline until the late nineteenth century, so the interest of psychologists in the moon illusion is largely associated with this view of perception. Psychologists from this school of thought usually explain the celestial illusion as a deduction based on a mistaken perception of distance.

The phases described above overlapped to some extent, with rival perceptual theories running in parallel. They continue to do so. Currently there is a revival of interest in 'direct' perception, generated by J.J. Gibson (1904–79). In this modern version of an ancient theory, stimulus information is regarded as sufficient for veridical perception. The followers of this theory have little to say about the moon illusion.

The extensive literature on the illusion is spread over many centuries in several languages. This has led to some historically interesting anomalies. For example, almost identical explanations of the moon illusion – in terms of the distance to the horizon over which the moon is seen – were proposed during the seventeenth century by Castelli[20] and also quite recently;[21] and one of the best-known accounts – the 'flattened dome' – is often attributed to Ptolemy, whereas it actually originated in medieval times.[22] We hope that this book will help to correct such historical inaccuracies.

Size of the enlargement

One of the most basic questions about the celestial illusion is its size: how much larger do the sun, moon and constellations on the horizon appear to be than when higher in the sky? This question has led to many different answers, which we will briefly review. At present we will discuss only estimates of the illusion under natural conditions, while attempts to measure it with the aid of comparison discs will be considered in Chapter 4.

Many observers have given only qualitative descriptions. For example, the German naturalist Alexander von Humboldt, when discussing the tropics, notes that 'some of our northern constellations – as for instance Ursa Major and Ursa Minor – owing to their low position when seen from the region of the equator, appear to be of a remark-able, almost fearful magnitude' (1850, p. 349). Similarly, the horizon moon appeared 'occasionally of an extraordinary large size', according to Walker (1806, p. 241). Such descriptions are of little use for understanding the illusion, so we will concentrate on more precise estimates.

An early quantitative estimate was given by Molyneux (1687), who implies a horizon enlargement of the moon of a factor of 3 or 4, but later adds that 'very often have I seen the Moon when she appeared 10 times broader than ordinary' (p. 322). Thus for him the illusion was quite variable, and sometimes very pronounced. This was supported by Logan (1736, p. 404) who reported an extraordinary enlargement, which he put at more than ten times when 'seeing the sun rise or set over a small eminence at the distance of a mile or two with tall trees on it'. Haenel (1909, p. 164) also agreed that the horizon enlargement of the moon amounts to '2-, 3-, yes 10 times its zenith size'. Robert Smith (1738, p. 64) found it to be about three and a third times, while Solhkhah and Orbach (1969, p. 88) also 'believe that a ratio even greater than 3 to 1 is not uncommonly experienced'.

Huyghens[23] more modestly claimed that 'the disc of the sun seems almost doubly greater at the horizon than when it is more elevated'. This estimate of a factor of about 2 for the apparent enlargement is often encountered. It was given by Malebranche for the sun,[24] and was found for the moon by Stroobant (1884). An enlargement of 'two or three times' was also reported by Hutton (1796, vol. 2, p. 73), while Dember and Uibe (1920, p. 353) say 'more than double'. These numerical estimates represent the impressions of only a small number of observers, and the values are much larger than the average estimates of about 1.5–2.0 given by classes of psychology students today. Other methods of measurement, which we discuss in Chapter 4, also give more modest enlargement values.

We should perhaps distinguish between the 'normal' illusion, with an enlargement of about 1.5, and the 'super' illusion with a larger factor. The authors of this book have experienced the super illusion only rarely. One of us (CP) has seen it twice, many years apart, when the dim and reddened sun set in a haze on a far horizon with much detail on it. On both occasions the illusion was compelling, excessive, and immediately perceived by each of the few persons present. The other (HER) experienced it while working on this book one evening: she looked out of the window across a flat plain to the nearby hills in the north-east, where the pale yellow moon was rising rapidly behind the lower slopes and appeared huge. Her first, illogical, thought was 'Ah! It must be the sun to appear so large.' A minute or two later the moon slipped away from the hillside, changed to a bright white colour, and shrank to a more normal size. When she mentioned this to a friend, he reported having seen the red sun setting behind a large cooling tower: it appeared like an enormous glowing ball in the sky, and for a moment he did not know what it was; he was driving at the time, and switched on the radio hoping for a news flash on this unidentified flying object. Another colleague wrote: 'I saw the moon on the horizon early in the morning – before daylight – while driving to work, and it took me several seconds before I realized what it was. My first thought was an illuminated hot-air balloon.' Such mistakes are, of course, not new. For example, Angell (1924) wrote that she 'once mistook the disk of the moon, seen through the top of a group of lofty poplars about a quarter of a mile distant, for the disk of a windmill placed among the trees. The disk seemed about 10 feet in diameter.' Clearly the super illusion is rare, but compelling, and often involves a cognitive mistake.

In summary, the apparent enlargement of the sun and moon on the horizon is often estimated to be about two times, but occasionally more. The horizon enlargement of the constellations, or of the distances between stars, is usually estimated to be less than this, as was noted by Bourdon (1899). Huyghens[25] remarked that 'very often the stars of the Northern Great Bear when they approach the horizon, will seem to be double the distance from each other'; but Stroobant (1884), who carefully compared the estimated distances between pairs of stars at various elevations, put the enlargement at only about 25 per cent. Yet Robert Smith (1738, p. 68) had previously used this method to confirm his estimated horizon enlargement of three and a third times. Backhouse (1891), using the same technique, found that for him the enlargement amounted to at most 10 per cent for small distances on the celestial sphere, and even less for larger separations. With pairs of stars which were actually 26 degrees apart no horizon enlargement was observed, while for still larger separations the illusion was reversed. Filehne (1894), on the other hand, found an enlargement of about two times with a similar comparison technique. Indow (1968) found no horizon enlargement at all, when a number of observers estimated the distances between pairs of equally bright stars at various elevations. Reed and Krupinski (1992) also found no illusion for star pairs, and a smaller illusion for star triads than for the moon, when a large number of observers made matches of angular size, and estimates of linear separation, for the same celestial targets at different elevations.

Why are the sun and moon illusions estimated at a factor of 2 or more, while the estimated enlargement of the distances between stars is usually much less? We will consider a methodological reason in Chapter 4, but one aspect can be pointed out now.

When estimating the enlargement of the sun or moon, its size at the time of observation must be compared to its remembered size at an earlier time, usually hours or even days ago. On the other hand, the distances between two pairs of stars at different elevations can be seen either simultaneously or in quick succession. These two methods of judgement involve different cognitive mechanisms. The comparison of two simultaneously visible distances on the celestial sphere is a relatively simple perceptual task; but the estimation of the horizon enlargement of the sun or moon is more complex as it involves memory. The method of estimation may affect the size of the illusion. We note that the pronounced enlargements reported by Huyghens and von Humbolt applied to whole constellations, rather than to distances between pairs of stars, and were thus based on comparisons with remembered sizes.

Variability of the illusion

The estimates reported above show that the horizon enlargement is not always the same.[26] Several factors may contribute to the variability. Obviously, the elevation of the celestial body at the time of observation must be one of these factors. The size of the illusion has been reported by some to decrease rapidly as the sun or moon rises above the horizon, almost disappearing at an elevation of a degree or two (Dunn 1762) or somewhat more (Müller 1906, 1907). Others report a much more gradual diminution with elevation (Eginitis 1898), while some report variable findings (Dember and Uibe 1918). Thus, estimates for horizon viewing could vary depending on whether the moon is right on the horizon, only partly visible above the horizon, or just clear of the horizon. Similarly, if the enlargement disappears only at the zenith, the full extent of the illusion cannot be experienced at high latitudes where the sun and moon are never seen directly overhead.

People differ in their susceptibility to the illusion, and their perception may change with time or experience. Some people with training in the natural sciences tend to scoff at illusions, denying that they are subject to them. They may claim that any experienced enlargement is 'real', i.e. that it can be measured with a sextant or similar instrument. But even among scientists denial of the illusion is the exception rather than the rule. 'The astronomer falls … into the same deception as the most ignorant clown' said Euler in 1762,[27] and his view was supported by Zeno (1862). Stroobant (1884, 1928), himself an astronomer, also experienced the illusion strongly. Most observers who studied the enlargement were aware of the fact that it was illusory; so knowledge or disbelief concerning the illusion does not usually lead to its disappearance. However, the opposite is occasionally claimed, as for example by Lewis (1898, p. 392):

> I may add that since my careful study of the subject I have almost entirely lost the illusion in the case of both sun and moon, but I have never been able to divest myself from it in the case of the apparent size of the constellations, or the distance apart of fixed stars when near the horizon as compared with their appearance at their meridian altitudes.

The different factors that may contribute to the variability of the illusion will be considered in later chapters, because they play key roles in its explanation. Meanwhile, before discussing the illusion any further, we shall establish some astronomical facts about the sun and moon.

Summary

Celestial bodies, such as the sun, moon, and constellations, appear larger near the horizon than on high. The illusion has been widely noted, and may have been described as early as the seventh century BC. The main early literary sources are from Greek, Latin, and Arabic. Interest in the illusion has moved through several different disciplines over the centuries: general science and philosophy, then astronomy, then physiological and meteorological optics, and finally visual psychology. The understanding of visual perception also developed: originally all illusions were assumed to be physical phenomena, but later the distinction between perceptions and reality emerged, and modern psychologists enquire how the brain constructs percepts from retinal stimulation. The celestial illusion varies in size with many factors. The sun and moon illusions have often been described as a factor of about 2, or occasionally much more, between the horizontal and elevated appearances. The constellation illusion is reported as much less than a factor of 2.

CHAPTER 2

The real sizes of the moon and sun

The simplest explanation of the horizon enlargement of the celestial bodies would be that it is real: in other words that the moon, sun, and constellations are measurably larger when rising or setting than at other times. The moon and sun are of course considered to be relatively unchanging physical objects; therefore the argument would have to be that their angular size is somehow enlarged while their linear size remains the same. Such an enlargement could be due to the celestial body being closer to the observer when rising or setting, and thus subtending a larger angle. We must therefore describe the variations in the distances of the celestial bodies and discuss whether these could partly explain the perceived horizon enlargement or its variability. First we need to define a few physical terms that are used in the measurement of the celestial bodies.

Physical measures of size and altitude

Real or physical size can mean various things, depending on the object in question and the purpose of measurement. We will speak of linear size when the object has a one-dimensional magnitude, usually expressed in metres: thus the diameter of the moon has a real linear size of some 3500 km. When there is more than one dimension, size refers to area or volume, usually expressed in square or cubic metres. The term *angular size* will be used for the one-dimensional angular extent of an object, usually expressed in degrees and minutes of arc. The angular diameter of the moon is about half a degree, meaning that the diameter of the moon subtends an arc of about half a degree at the eye of the observer (Fig. 2.1). Astronomers have long used the term *apparent size* for what we call *real angular size*, but this usage is clearly to be avoided when dealing with illusions.

For objects of small angular size such as the sun and moon the relationship between angular size (A), linear size (L) and distance (D) is approximately: $A = L/D$, when A is expressed in radians (Fig. 2.2). If any two of A, L, and D are known, the third can be calculated.

When describing how high the moon is in the sky, astronomers use the term *altitude*, which signifies the angular distance of the centre of the moon's disc perpendicularly above the horizontal plane through the eye of the observer. However, this term has other meanings outside astronomy and we therefore prefer to speak of the *angle of elevation* of the moon.

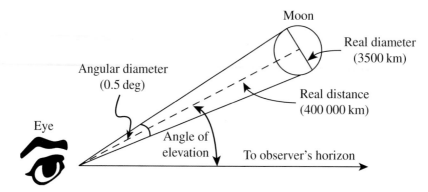

Figure 2.1 Some terms to describe the real size, distance and position of the moon. The numerical values are approximate.

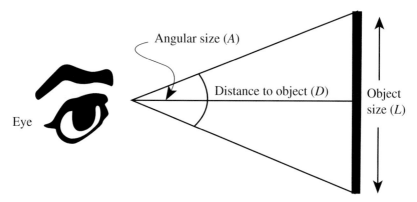

Figure 2.2 The angular size (A) of an object is related to its linear size (L, measured at right angles to the line of sight) and its distance (D) by: tangent $(A/2) = L/2D$. However, for small angles the tangent of an angle is approximately equal to the angle itself, when the latter is expressed in radians. For objects of small angular size the relationship therefore reduces to approximately: $A = L/D$.

The term *zenith*, meaning the point on the celestial sphere vertically above the observer, has been very loosely employed by writers on the moon illusion to indicate any position some distance above the horizon, or sometimes the position of the sun or moon when at their maximum angle of elevation on a particular day. As the sun and moon are never seen directly overhead in higher latitudes the term zenith is seldom appropriate when describing the celestial illusion.

The angular sizes of the moon and sun

The moon is a roughly spherical, solid body which is visible to the unaided eye when it is directly illuminated by the sun. It moves around the earth in a slightly irregular elliptical orbit. Together with the earth it moves around the sun in the course of a year, the mean distance from the earth to the sun being about 400 times larger than the mean distance to the moon. Using the line from the earth to the sun as a reference direction, the moon completes an orbit round the earth in about 29 ½ days. This is consequently the period between one full moon and the next (Fig. 2.3). The moon appears full when it is opposite the sun, so the full moon rises at about the time when the sun is setting (at least in lower latitudes), and reaches its maximum elevation near midnight. During the few days when the moon is nearly full it can therefore be observed to rise in daylight (before full moon), during twilight (at full moon), and in the dark (after full moon).

The mean angular diameter of the moon is about 31.1 minutes of arc,[1] but its distance from the centre of the earth is not constant. Its orbit round the earth is an approximate ellipse, whose orientation changes only slowly with reference to the fixed stars: the moon will therefore be seen as full at various points in this orbit, and thus at various distances, in the course of a year. As a result its angular diameter changes slightly from one full moon to the next. It is sometimes as small as 29.4 minutes of arc and at other times as large as 32.8 minutes of arc. This difference of 3.4 minutes of arc represents a variation in angular size of about 11 per cent of the mean value (or 5.5 per cent above and below the mean). A difference of about 5 per cent is easily detectable for line lengths presented simultaneously, since the detection threshold for this task is known to be as fine as 1 per cent[2] or 2 per cent.[3] Differences in the diameter of a circle are probably even easier to detect, since the area of a circle increases with the square of the radius. However, gradual variations in the size of isolated objects viewed at different times are much less obvious. On the whole, it seems unlikely that even the

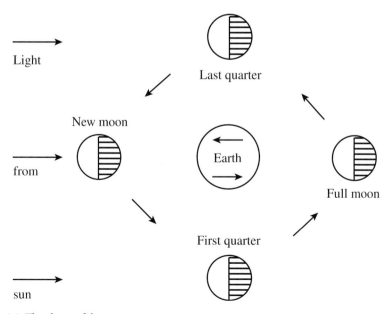

Figure 2.3 The phases of the moon.

most extreme variations in the angular size of the moon would be perceptible with normal observations.

It remains possible that such variations might contribute to the variability of the illusion. During the eighteenth century Desaguiliers[4] claimed that 'the difference of distance of the moon in perigeo [closest point] and apogeo [furthest point], will account for the different bigness of the horizontal moon at different times'. More recently this proposal was revived by Osaka (1962). He gave details of the earth–moon distance for the preceding year, and noted that the difference between the largest and smallest angular diameters of the moon might be detectable by the unaided eye if the two images could be seen side by side. This is indeed the case, as is evident from the juxtaposed photographs of a near and far moon shown in Fig. 2.4. Osaka therefore advised that 'experimental psychologists who are going to engage in the measurement of the moon illusion should not forget monthly variation of the astronomical distance between the moon and the earth' (1962, p. 28). However, in terms of the large variability of the illusion mentioned earlier, these variations in angular size are very small.

Full moons at certain times of the year have sometimes been given specific names. The best known of these is the harvest moon, which is the full moon nearest to the time of the autumn equinox in the northern hemisphere (22 September). The moon's path on the celestial sphere is such that at this time, and at the latitude of Britain, the full moon rises only slightly later each evening for several days. The name harvest moon indicates how useful a few evenings of moonlight were found to be for agricultural purposes. The phenomenon occurs again, though to a lesser extent, for the next full moon (in October), which is sometimes called the hunter's moon. The popular belief that the harvest moon appears particularly large when rising cannot be justified on astronomical grounds. The date at which the full moon is closest to the earth changes from year to year over a cycle of about 18 years: the largest and smallest full moons are therefore not linked to any particular season. The apparent enlargement may simply

Figure 2.4 Photographs of the moon when closest to the earth (left) and furthest from the earth (right). The difference in diameter is about 12 per cent. From Walker (1997).

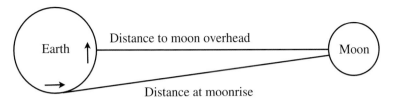

Figure 2.5 The moon is closer when high in the sky than when it rises or sets. Its mean distance is about 60 times the radius of the earth.

reflect the fact that the harvest moon receives more attention than do full moons at other times of the year. Alternatively, autumn mists may enhance the contribution of aerial perspective (Chapter 6).

A second factor that influences the distance between the moon and an observer on the surface of the earth is the daily rotation of the earth on its axis. This rotation carries the observer closer to the moon as it rises higher in the sky, and further away again as it sets (Fig. 2.5). The change in distance is largest for an observer at the equator. At this position the observer is furthest from the earth's axis, and the overhead moon is closer than the rising moon by the length of the earth's radius (6378 km). The mean distance to the moon is about 60 times the radius of the earth, so the corresponding increase in the angular size of the moon as it rises from the horizon to the zenith – known as *augmentation* in astronomy – is only half a minute of arc on average. This is much smaller than the variation of 3.4 minutes of arc occurring during a year, and is probably below the threshold of discrimination. In higher latitudes, where the observer is closer to the earth's axis, the effect is even less. Its interest therefore lies mainly in the fact that the moon is actually slightly smaller in angular size near the horizon than at higher angles of elevation, contrary to its perceived horizon enlargement.

The angular diameter of the sun varies even less than that of the moon, as the distance of a terrestrial observer from the sun is relatively constant. The earth's orbit around the sun is not quite circular, but rather an ellipse with the sun in one focus. The point of closest approach to the sun (called *perihelion*) is reached early in January and the sun then has its maximum angular diameter of 32.5 minutes of arc. In early July the distance is greatest (at *aphelion*) and the sun then subtends its smallest angular diameter, namely 31.5 minutes of arc.[5] Its variation in the course of a year is too small to play any significant role in the variability of the sun illusion.

The mean distance of the sun to the earth is almost 24 000 times the radius of the earth. The change in angular size of the sun due to the rotation of the earth is therefore much less than that of the moon, and is completely insignificant for our purpose.

Solar and lunar eclipses

A solar eclipse occurs when the moon is directly between an observer on earth and the sun, thus hiding the sun from view (Fig. 2.6). This can happen only at new moon.

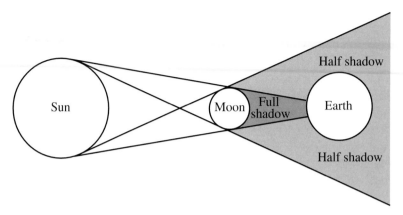

Figure 2.6 Schematic representation (not to scale) of the positions of the sun, moon, and earth at the time of a solar eclipse. On the small portion of the earth's surface in the full shadow of the moon a total eclipse is visible. Observers in the partial shadow of the moon will see a partial eclipse.

The moon's orbit around the earth is slightly inclined towards the plane of the earth's orbit around the sun, with the result that eclipses do not occur at every new moon, but only when all three bodies are in a straight line. Furthermore, most solar eclipses are partial rather than total, in the sense that the moon obscures only a portion of the sun's disc. If the angular size of the moon were much larger than that of the sun, total eclipses would be fairly common. On the other hand, if the angular size of the moon were smaller than that of the sun, total solar eclipses would be impossible because the moon would be too small to cover the whole disc of the sun. By a rare coincidence, the angular sizes of the moon and sun are approximately the same, as explained in the previous section. As a result, total eclipses do occur, but are rare, and each one is visible from only a small portion of the earth's surface. They are spectacular events, which relatively few people are privileged to witness.

A lunar eclipse occurs when the earth is directly between the moon and the sun, so that the moon is in the earth's full shadow (Fig. 2.7). This can happen only at full moon. For a (hypothetical) observer on the moon, the angular size of the earth is much larger than that of the sun. Hence total lunar eclipses occur more frequently than total solar eclipses. Lunar eclipses can be very beautiful, with the shaded part of the moon appearing reddish because the light that reaches it has been refracted through the atmosphere (Chapter 5). The shadow also gives the moon depth, so that it appears like a solid globe rather than a flat disc.

Both lunar and solar eclipses have been observed and recorded for many centuries. Ptolemy (second century AD) used the eclipse observations of Late Babylonian astronomers, combined with his own observations, to calculate the elements of the moon's orbit. The earliest of these Babylonian observations was a total lunar eclipse observed in Babylon in 721 BC.[6] The causes of both solar and lunar eclipses were clearly understood by astronomers even before Ptolemy's time.

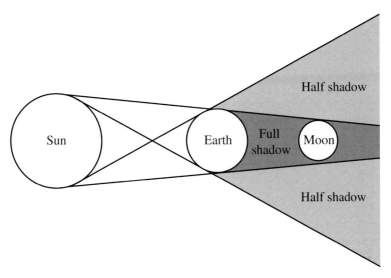

Figure 2.7 Schematic representation (not to scale) of the positions of the sun, earth, and moon at the time of a lunar eclipse. If, as shown here, the moon is completely immersed in the full shadow of the earth, the eclipse is total.

Angular size and the moon illusion

The insignificance for the celestial illusion of the changes in the real angular size of the celestial bodies has been accepted by scientists since antiquity. The Greek astronomer Hipparchus (second century BC), and later Ptolemy, took the sun's angular diameter to be constant and equal to the 650th part of a full circle, or about 33 minutes of arc.[7] Ptolemy put the moon's mean distance at 59 times the radius of the earth, which is quite accurate. Although his theory of the moon's orbit overestimated the variation in the earth–moon distance, he nevertheless regarded this variation as being fairly small, giving rise to only slight variations in the moon's angular diameter.[8] In fact, he stated quite clearly that it was not their closeness that made the celestial bodies appear larger on the horizon.[9] Cleomedes (*c.* early third century AD) similarly concluded that 'It is necessary to consider all effects of this type as phenomena affecting our sight, and certainly not as properties of the objects we see.'[10]

Astronomers have always attached particular importance to the angular diameters of the moon and sun, as these are required for the successful prediction of eclipses. It is therefore not surprising to find that the diameters have often been measured since antiquity.[11] The accuracy of such measurements increased during the first half of the fourteenth century when Levi ben Gerson (1288–1344) improved and popularized the camera obscura. This consisted of a dark box with a tiny aperture in one side, by means of which an image of the sun or moon could be formed on the side of the box opposite the aperture. Having been developed by Ibn al-Haytham three centuries earlier, it was used by Levi to study the small variations in the angular diameters of the sun and moon.[12] Further improvements in accuracy became possible when telescopes fitted with micrometers came into use during the seventeenth century. It has thus been known for a considerable time that there is a minor variation in the moon's size due to the rotation of the earth, and that it is contrary to the perceived horizon enlargement. Leonardo da Vinci[13] was aware of this fact in the beginning of the sixteenth century, and the French philosopher Nicolas Malebranche[14] referred to it in 1675, which shows that such knowledge was not confined to astronomers. It was mentioned also by Molyneux (1687) and Berkeley (1709), while Euler explained it carefully in one of his letters of 1762.[15]

Even in antiquity, scientists never seriously considered the horizon enlargement of the sun and moon to be caused by their being closer than when on high. However, there is a non-scientific source that contains this idea, and which is of interest because of its early date and unique point of view. The source is a collection of Chinese legends and folk-tales ascribed to Lieh Tzu, and may date from any time between the fourth and first centuries BC. It contains an apocryphal story about the philosopher Confucius (c. 551–479 BC) meeting two boys who were arguing about the distance of the sun (Giles 1923, pp. 68–9). An English translation[16] of the passage is as follows:

> When Confucius was travelling in the east, he came upon two boys who were disputing, and he asked them why. One said 'I believe that the rising sun is nearer to us and that the midday sun is further away.' The other said 'On the contrary, I believe that the rising and setting sun is further away from us, and that at midday it is nearest.' The first replied 'The rising sun is as big as a chariot-roof, while at midday the sun is no bigger than a plate. That which is large must be near us, while that which is small must be further away.' But the second said 'At dawn the sun is cool but at midday it burns, and the hotter it gets the nearer it must be to us.' Confucius was unable to solve their problem. So the two boys laughed him to scorn saying, 'Why do people pretend that you are so learned?'

It would appear from this story that the sun illusion was well known in China at the time of writing. We also note that it was the real angular size of the sun that was thought to be enlarged on the horizon, and this in turn was taken to imply that the sun must be closer at that time.

Around this time informed opinion in China already favoured a terrestrial rather than an astronomical explanation of the illusion. For example, during the Han dynasty (from 206 BC), Chang Heng described the illusion as an optical phenomenon. A few centuries later, in the late third or fourth centuries AD, Shu Hsi also stated that the sun does not change in (real angular) size, and that its apparent size changes were the result of a deception of our senses.[17]

Although the distance theory was not seriously considered by astronomers, there is a brief passage in Ptolemy's *Almagest* (pp. 7–8) which suggests that such an explanation would not have been thought far-fetched at the time. Ptolemy found it necessary to criticize the opinion that the celestial bodies moved in a straight line to an infinite distance when setting, being created anew each day. This idea was credited to the Greek philosopher Xenophanes (c. 570–470 BC), who expressed it in an obscure passage.[18] As an argument against this view Ptolemy asks: 'Or how is it they do not disappear with their size gradually diminishing, but on the contrary seem larger when they are about to disappear, being covered little by little as if cut off by the earth's surface?' The argument clearly implies that Ptolemy thought the enlargement to apply to real angular size, although, as we saw above, he did not attribute this to greater proximity.

The above passage is of particular interest because it appears to have come to the attention of Chinese astronomers at an early date. The astronomical chapters of the official history of the Chin dynasty (265–420), written in the seventh century under the supervision of Emperor T'ai Tsung, contain a discussion by the scholar Ko Hung (283–343) of the theory that the sun cannot be seen at night because it moves a great distance away from the earth. Ko Hung then argues: 'If it is said that the sun can no longer be seen because it has gone a very great distance from us, then when the sun sets, on its way towards the north, its size should diminish. But, on the contrary, the sun becomes larger at sunset. This is far from being proof that the sun is moving further away when it sets.'[19] The argument is very similar to Ptolemy's argument of a century or two earlier.

The similarity may have arisen because Chinese conquests in central Asia in the first century AD brought about communication with Iran and led to regular overland trading with the Roman empire.[20] Along these routes Greek science travelled via India to China. In fact there are references in Chinese sources to persons acquainted with astronomy who travelled to China from the Middle East in 164. Ptolemy's books on astronomy, written about 142, could have reached China in this manner.

The angular sizes of the stars

The fixed stars are at such large distances from us that their angular sizes are extremely small. Even through the largest earth-bound telescopes the stars appear as mere points of light with no measurable angular size. Yet to the unaided eye the bright stars and planets appear larger than mere points. The enlargement is mainly the result of imperfect focusing by the lens of the eye due to spherical aberration. Light rays passing through the outer portions of the lens are brought to a slightly shorter focus than light rays passing through its central parts. As a result a point source of light such as a star cannot be brought into sharp focus when the pupil of the eye is wide open at night. Bright stars appear to be larger than dim ones because a larger part of their out-of-focus image on the retina is bright enough to excite the retinal receptors.[21] This phenomenon convinced ancient astronomers that Venus, the brightest planet, had a mean angular diameter of 3 minutes of arc, or about one-tenth of the angular diameter of the sun.[22] Its true angular diameter is, however, always less than about 1 minute of arc. Viewing the stars through a small artificial pupil reduces their perceived size – a phenomenon that much impressed Leonardo da Vinci:[23] 'If you look at the stars, cutting off the rays (as may be done by looking through a very small hole made with the extreme point of a very fine needle, placed so as almost to touch the eye), you will see those stars so minute that it would seem as though nothing could be smaller.' They also look smaller during twilight, when there is more light and the pupil is therefore smaller. This phenomenon was used by Galileo to prove that the perceived size of the stars was an illusion and that their real angular size was too small to be measured.[24]

We are aware of only one case where a scientist expressed the belief that the celestial illusion might also affect individual fixed stars. During the reign of al-Malik al-Kamil, sultan of Egypt from 1218 to 1238, King Frederick II of Sicily sent him a number of difficult problems to test the learning of Muslim scholars. Frederick was not only a prominent supporter of scientific pursuits, but was himself an intellectual of note and the author of an outstanding treatise on falconry.[25] Three of his questions to the sultan have been preserved: why do objects partly covered by water appear bent?; what is the cause of the illusion of spots before the eyes?; and (the one that mainly concerns us), why does Canopus appear bigger when near the horizon whereas the absence of moisture in the southern deserts precludes moisture as an explanation?[26] These questions are all examples of the fallibility of the senses, a problem that had been discussed by many earlier philosophers and scientists.[27] The bent stick in water was a stock question, but the Canopus question is unusual.

Canopus is the second-brightest fixed star, situated deep in the southern celestial hemisphere. From southern Sicily it might be just briefly visible very low over the southern horizon in winter. However, Frederick's reference to the 'southern deserts' shows that he had observations from lower latitudes in mind, and it is known that he personally visited Syria.[28] His rejection of moisture in the air as the cause of the horizon enlargement will be clarified by the discussion of atmospheric theories of the illusion in Chapter 5.

Summary

The sun and moon both have an angular size of about half a degree, when viewed from the earth. In the course of a year the moon varies in its distance from the earth, causing a variation of up to 11 per cent in its angular size. This variation is not a cause of the horizon enlargement, but could add to the variability of size estimates at different times. The earth's diurnal rotation causes the moon to be nearer, and subtend a larger angle, when overhead than on the horizon; but the difference is too small to be detectable, and is contrary to the perceived horizon enlargement. Variations in the angular size of the sun are even smaller than for the moon. The perceived size of the stars is much greater than their angular size, which is too small to be measured.

CHAPTER 3

Perceiving size

We have been using terms like *apparent size* rather loosely up to now. Unfortunately such looseness is common in studies of the moon illusion, and is responsible for some of the differences of opinion about its cause. As recently as 1989, in Hershenson's edited book on the illusion, the contributors used many different expressions for the same concept, and the same expression for several different concepts. If the moon illusion is to be understood, it is first necessary to clarify the nature of real and apparent size and to use an agreed terminology. In the first place we need to distinguish unambiguously between a size as found by objective physical measurement, and a subjective impression of size in a human observer. We will usually refer to the former as real size whenever confusion is possible. Actual, true, or physical size can be regarded as synonyms. We have already defined some physical terms in the previous chapter – those relating to linear size and angular size. We now turn to the more controversial issue of subjective size.

Subjective measures of size

We need a term for the subjective impression of largeness that an observer has when attending to the size of an object. The term *perceived size* seems appropriate, though *phenomenal size* or *apparent size* are often used synonymously. Apparent size is the term most often used by psychologists, but is best avoided in general science literature because astronomers and physicists use it to mean true angular size.[1] The perceived size of an object is something private to the observer, but in order to study it some numerical expression is needed. Such numbers can be obtained in several ways, and it is wise to use distinct terms for them.

Firstly, the observer can be asked to provide a numerical size estimate, or *estimated size*. These estimates may be expressed in customary units of measurement, such as metres for linear size and degrees for angular size; alternatively, in the method known as *magnitude estimation*, arbitrary units may be used.[2] Secondly, the observer can be asked to make a match between the perceived size of the object in question and that of some other comparison object; in this method the numerical values are obtained by the experimenter, who measures the matched comparison object. We shall refer to this second type of judgement as a *matched* size. Matching methods are more commonly used in measurements of the celestial illusion, while both numerical estimates and size matches are used for terrestrial objects.

It is not always clear whether numerical estimates relate to the real size or the perceived size of an object. An estimate such as 'the tree seems about five metres high' means that the observer believes that if the height of the tree were measured it would be found to be approximately five metres. This is therefore an estimate of the real linear size of the tree. On the other hand a statement such as 'the moon seems about as large as a balloon' is quite different, reflecting the perceived size of the moon. Size estimates can reflect either real or perceived size depending on the instructions.[3] In the case of the moon this particular ambiguity cannot arise: the moon appears to be at an indeterminate distance and we are unable to estimate its real size.

There are two other sources of ambiguity that must be considered. First, does the estimated size of the moon relate to its diameter or to the area of its disc? When an observer reports that the rising moon appears to be 'twice as large' as usual, does he mean that its diameter, its area, or some unknown combination of these is judged to be doubled? When you glance quickly at Fig. 3.1, how many times bigger does the large circle seem than the small one? Is your perceived size more closely related to linear size or to area? There may well be individual differences in what people take the word 'size' to mean in such perceptual tasks. Referring back to the estimates of the magnitude of the celestial illusion (Chapter 1) we find that those of Molyneux, Logan, and some others relate explicitly to the perceived diameter of the moon or sun. On the other hand the French philosopher Malebranche, describing the sun when seen through a dark glass as 'about twice as near, and about four times as small',[4] was clearly referring to the area of the disc. When Huyghens estimated that the 'disc of the sun' seemed doubled near the horizon, he may have meant either that its diameter appeared to be doubled (and its area therefore more than doubled) or that its area appeared twice as large. Whichever of these we accept, the enlargement of the other measure cannot be found simply by calculation: the relation between the estimated areas and diameters is much more complex and variable than the geometrical relation between the real areas

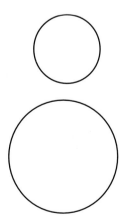

Figure 3.1 How many times was the smaller circle enlarged to produce the bigger one? The linear enlargement is 1.8 times, whereas the enlargement in area is 3.3 times.

and diameters would suggest.[5] We shall assume that most estimates refer to the perceived diameter.

A second and more intractable problem is whether the perceived size of the moon is analogous to linear or to angular size. Many people describe their perceived size of the moon in terms of a length, which suggests that they perceive a linear enlargement. However, many people also believe the enlargement to be real (and thus visible in photographs), which suggests that they perceive an angular enlargement. This dilemma affects both experimental method and theory: an investigator's assumption about the nature of perceived size determines both the choice of a measurement technique and the choice of a theory to explain the resulting measurements.

Many investigators have assumed that the perceived size of an object is similar to its real angular size. Others assume that perceived size corresponds to the linear size of the retinal image: this size is directly related to the angular size except for some minor distortions caused by the optics of the eye (Fig. 3.2). In these terms, the sun and moon have an angular size of about half a degree, which corresponds to a linear image size of about 0.15 mm for a standard eye focused at infinity. Some other authors suggest that perceived size is the conscious awareness of the size of the retinal image: this size is related to the real image size, but may be transformed by neural processes within the brain. These views all imply that perceived size is like angular size, and should be estimated in degrees or compared with some object of known angular size. They also imply that the moon illusion should be described as an increase in the real or perceived angular size of the moon near the horizon.

Other investigators have defined perceived size in terms of length rather than angle. According to this view, perception involves the conscious location of objects in a three-dimensional space, the latter being somehow constructed from our visual sensations by automatic thought processes. The perceived size of an object corresponds more or less to its size 'out there' – that is, to its real linear size – and can therefore best be expressed in metres. On this view the moon illusion should be described as an increase in the perceived linear size of the moon near the horizon.

These two assumptions about the nature of perceived size differ quite fundamentally, yet they have often been applied indiscriminately in the same investigation, with

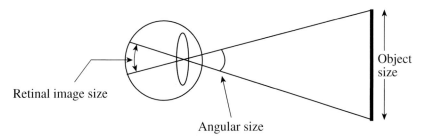

Figure 3.2 The retinal image size of an object varies systematically with the object's angular size, provided that the lens is correctly accommodated. If the lens is improperly focused, the image is blurred and slightly larger.

very confusing results.[6] Uncertainty about the nature of perceived size shows up in the statements and estimates that people make about the perceived linear size of the sun and moon. We describe some of these below.

How large do the sun and moon look?

Many investigators have asked the seemingly simple question 'How large does the moon appear to be?', and have obtained answers like 'as large as a plate' or 'about a foot across'. Some people – for example, astronomers and others who are used to thinking in terms of angular rather than linear distances on the celestial sphere – find the question mean-ingless and cannot or will not answer it.[7] Most people, however, do have impressions of the size of the sun and moon, and even the constellations.

Aristotle (384–322 BC) stated that 'the sun presents itself as only a foot in diameter'.[8] This shows that such estimates date back a long time. In fact, the philosopher Heraclitus of Ephesus (c. 540–480 BC) is said to have made a similar statement, namely that 'the breadth of the sun is that of a human foot'.[9]

Ancient astronomers often estimated the sizes of celestial bodies and distances on the celestial sphere in length units,[10] and many later observers have done the same. Thus the earliest known observation of a sunspot, recorded in China in 28 BC, describes it as being 'as big as a coin'.[11] In the early Middle Ages Isidore of Seville thought the sun appeared to be about a Roman cubit (444 mm) in width,[12] while Leonardo da Vinci (1452–1519) estimated it at about a foot.[13] Wilkins (1638, pp. 19–20) stated that com-mon people believed the moon to be the size of a cart wheel: 'You may as soon perswade some Country peasants, that the Moone is made of greene Cheese (as wee say) as that 'tis bigger than his Cart-wheele, since both seeme equally to contradict his sight, and hee has not reason enough to leade him farther than his senses.' About the same time Descartes (1637a) described the sun and moon as usually appearing to have a diameter of one or two feet; while for Porterfield (1759, Vol. 2, p. 373) 'the sun and moon are only circular planes of about a foot in diameter, if we believe the testimony of our eyes'. Hobbes (1658, Vol. 1, p. 75) also put the sun's size at one foot across, while Molyneux (1687) thought this to be about the size of the moon at high elevations, rising to three or four feet when close to the horizon. 'The sun and moon, each subtending about half a degree, appear in the meridian of the breadth of eight or ten inches … and in the horizon to be two or three foot' said Logan (1736, p. 404). Le Conte (1881, p. 159) stated that estimates differed: 'Where there are no means of judging of distance, we can not estimate size, and different persons will estimate differently. Thus, the sun or moon seems to some persons the size of a saucer, to some that of a dinner-plate, and to some that of the head of a barrel.' Most people questioned by Bourdon (1897, 1898) estimated the diameter of the moon when high 'without hesitation' at 200 to 300 millimetres. Similarly, many people compare the size of the moon to that of a plate, said Mayr (1904, p. 411) – an estimate which is supported by Witte (1918) as applying to the 'average' person under 'usual' conditions. A large plate, or a small table top, is also a common estimate for the sun and moon in southern Nigeria.[14] A football is another traditional object of about the right size (Fig. 3.3).

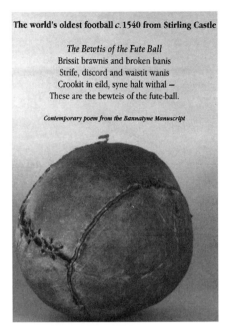

The world's oldest football *c.* 1540 from Stirling Castle

The Bewtis of the Fute Ball
Brissit brawnis and broken banis
Strife, discord and waistit wanis
Crookit in eild, syne halt withal —
These are the bewteis of the fute-ball.

Contemporary poem from the Bannatyne Manuscript

Figure 3.3 An old football, courtesy of the Stirling Smith Art Gallery and Museum. Football was a very rough game, as illustrated in the contemporary Scots poem, which translates as follows:

> Bruised muscles and broken bones
> Strife, discord and empty dwellings
> Crooked in age, then lame also –
> These are the beauties of the football.

Witte (1918) made some more systematic measurements, using three methods. First, he asked his observers 'How large does the moon seem to be?' and sought answers in terms of everyday objects such as a plate or one's head. Second, he required them to choose a paper disc of the same perceived size as the moon from a series with diameters of 10 to 500 mm. Third, he asked them to draw the moon on a large sheet of paper as the size it appeared to them. He used hundreds of observers, and found a mean value of about 200 mm with all three methods. Filehne (1917) also asked his observers to compare the moderately elevated full moon to paper discs, and reported that they selected a disc of about 200 mm (even when they protested against this method). The matched size at the highest elevations used was between 100 and 120 mm.

Despite these fairly consistent average results, considerable individual differences have been found. The observers questioned by Henning (1919) reported sizes from 5 mm to one metre. Angell (1932) for some years asked a large class of psychology students to estimate the size of the rising moon or sun, and obtained values varying from an American 25 cent piece (24 mm) to the size of a cart wheel.

From a geometric point of view these estimates make little sense, and consequently the scientifically minded find them very difficult to accept. They cannot be estimates of the real linear size of the moon (which has a diameter of about 3500 km) or the sun (with a diameter of about 1 300 000 km). The angular diameter of both the sun and the moon is about half a degree: to appear 200 mm across they should appear to be at a distance of about 24 metres, a distance much closer than the same observers usually estimate for the celestial bodies (Chapter 9). Perhaps the observers refer their size estimates to some close distance at which the chosen object or estimated linear size appears to subtend the same angle as the sun or moon.[15]

Does the tendency to ascribe some estimated size to the moon bear any relation to the moon illusion? Several authors have argued that it does: the moon must have some particular size, or it could not appear to be enlarged on the horizon.[16] With this question in mind, we will now consider some ideas on size perception that were held in classical times.

Early ideas on size perception

Philosophers and scientists in classical times were interested in how an image gets into the eye, and (to a lesser extent) how the image size is scaled to give a perceived size. These questions are partly interrelated because, on some theories, the image is scaled before or as it enters the eye. Greek theories of vision agreed that there must be physical contact between the eye and some aspect of the object of vision. Theories were *intromissionist, extramissionist,* or mixed. Intromissionists (such as the Epicureans) held that emanations or rays from the object entered the eye, and then produced a sensation. Extramissionists (such as Plato) held that rays or flux emanated from the eye and touched the object, and that the rays themselves were sensitive. Many philosophers (such as the Stoics) held mixed theories, in that rays or flux emanated from the eye and returned by the same route after interacting with aspects of the object in some manner. There was no necessary relation between belief in intromission or extramission and beliefs about size perception. Some authors believed that size was normally perceived correctly, some that it appeared to decrease with distance, and some that it varied according to several factors.

In early times an influential view was that of the atomists, who believed that perception was directly impressed upon the senses by intromission. The atomists included Democritus (*c.* 420 BC), Epicurus (*c.* 341–270 BC), and later the Roman poet Lucretius (*c.* 95–55 BC). According to Lucretius, the Epicureans believed that images were given off by objects, which were exact replicas of the shape, size, colour, and other properties of the object. These images were hollow films of very fine texture, which reached the 'soul' (brain) through various sense organs. If the images retained their correct size, object sizes should be correctly perceived.[17] However, according to Epicurean physics, an object at a distance appears smaller than close to, because the image gets worn down by other atomic bodies: thus the Epicureans did allow a distinction between true and perceived size for distant objects. Epicurus made an exception in the case of the sun, which he took to be about the size of a foot. Lucretius explained the reason for this as follows:

> Next, as to the size of the sun's blazing disc: it cannot in fact be either much larger or much
> smaller than it appears to our senses. So long as fires are near enough both to transmit
> their light and to breathe a warm blast upon our bodies, the bulk of their flames suffers no
> loss through distance: the fire is not visibly diminished. Since, therefore, the heat of the
> sun and the light it gives off travel all the way to our senses and illumine all they touch, its
> shape and size also must appear as they really are, with virtually no room for any lessening
> or enlargement.[18]

What Lucretius seems to be saying is that, when one observes a fire, it looks the same
size for any distance at which its light and heat can be sensed. The sun is similar to a fire,
since we can sense both its light and heat: therefore if it appears a certain size to us, that
must be more or less its actual size. Lucretius then applies a similar argument to the
moon: size appears constant over all distances at which an object is clearly visible; since
the moon has a sharp outline and clearly perceived shape, it too must be more or less as
large as it appears.

Most later philosophers were not swayed by these arguments, and were not prepared
to place as much trust in the evidence of their senses as Epicurus and Lucretius did.
The Epicurean position was attacked by many scholars, particularly the Stoics, most
of whom held extramission or mixed theories. One was Cleomedes, a Greek astronomer
of about the early third century AD,[19] who pointed out that the celestial illusion implied
that we could not always see the sun as its true size. Another critic was Plotinus
(*c.* AD 205–70), who, though he believed in incoming rays, argued that the soul
did not receive impressions of objects: if it did, it would not be able to perceive distance,
since the impression was no distance away; nor size, since objects such as the sky
were too large to fit inside the soul. Therefore, the soul must receive a translation
of an impression.[20]

Many scholars were aware that perceived size could change with distance and
that errors of size perception could be made, and they speculated on how size
was scaled. The idea that objects are correctly scaled despite changes in distance
came to be known as *size constancy*.[21] Two main lines of thought developed.
One was concerned with *relative size*, and the other with *taking account of
distance*.

The idea of relative size was expressed by Plotinus, who argued that size scaling
was given by the details within objects, and that size constancy fails at great distances,
not because the angular size is small but because the details are lost. He wrote
(*Second Ennead*, VIII):

> But the phenomenon [of size scaling] is more easily explained by the example of things of
> wide variety. Take mountains dotted with houses, woods and other land-marks; the obser-
> vation of each detail gives us the means of calculating, by the single objects noted, the total
> extent covered: but, where no such detail of form reaches us, our vision, which deals with
> detail, has not the means towards the knowledge of the whole by measurement of any one
> clearly discerned magnitude ... It was the detail that prevented a near object deceiving our

sense of magnitude: in the case of the distant object, because the eye does not pass stage by stage through the stretch of intervening space so as to note its forms, therefore it cannot report the magnitude of that space. The explanation by lesser angle of vision has been elsewhere dismissed.[22]

The other main explanation of size constancy is often specified as the *size–distance invariance hypothesis*.[23] This hypothesis states that the ratio of an object's perceived linear size to its perceived distance is determined by the ratio of its real linear size to its real distance: this ratio in turn depends only on the object's real angular size. We now consider the development of this hypothesis, which became the dominant account of size constancy.

The development of size–distance invariance

Early mathematical scholars (such as Euclid, Ptolemy, and Cleomedes) usually described their ideas in terms of visual rays or a visual cone, which emanated from the eye and touched the object. Sensation was achieved either because the rays themselves were sensitive, or because the rays returned along the same route bringing back an image to the sensitive eye. While the theory of outgoing rays caused difficulties for the understanding of sensation, it did not affect the application of geometrical reasoning to perception. Several centuries passed before the Arab mathematician and physicist Ibn al-Haytham (*c.* 1040) clearly stated that the direction of the rays was incoming, from the object to the eye.

Euclid, who lived around 300 BC, generally equated visual angle with perceived size. He wrote in his *Optics* (Theorem 5): 'Objects of equal size unequally distant appear unequal and the one lying nearer to the eye always appears larger.'[24] In this passage Euclid uses the language of *appearances*. In another passage (Theorem 21 – 'To know how great is a given length') he argued that linear size could be calculated in a geometrical manner from the angular size and the distance: but in this passage he uses the language of *calculation* rather than appearances.[25]

A different view was that linear size was *perceived* rather than calculated. One of the earliest statements of this view was made by Ptolemy in his *Optics* in the second century AD.[26] He based this work partly on Euclid's *Optics*, and probably on other books that are now lost. He included a long discussion of both size and shape constancy.[27] A typical passage (*Optics*, II, 56) reads:

> Suppose there are two lengths, AB and GD, which have the same inclination and subtend the same angle, E. When the distance of AB is not equal to that of GD, but shorter than it, AB will never appear longer than GD when determined by its true distance. Instead it will appear either shorter, which would happen when the difference in distance of the two lines is distinguishable; or equal, if the difference in distance is indistinguishable (the letters refer to Fig. 3.4).[28]

The writings of Cleomedes contain a rather clearer statement that the same angular size gives rise to different perceived linear sizes at different perceived distances.[29] In his book *On the circular motion of the heavenly bodies* (Book II, Chapter 1, section 4)

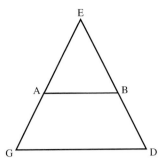

Figure 3.4 Ptolemy's diagram for size–distance invariance (*Optics*, II, 56).

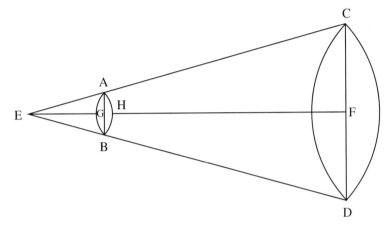

Figure 3.5 Balfour's diagram (1605) to illustrate Cleomedes' idea of size–distance invariance.

he wrote:

> The sun seems to us sometimes larger and sometimes smaller, with its distance corre-
> spondingly greater or smaller. A cone is formed by the rays that our sight emits, and if the
> cone truly reaches the sun it must be very large. Since the size and distance of the sun can
> also become very small in appearance, we can conceive of two cones, one reaching the sun
> in reality, the other in appearance. The apex of these cones is the same – the point situated
> at the pupil of the eye; but the base of one corresponds to reality and the base of the other
> to appearance. Consequently the relation between the real and apparent distance is the
> same as that between the real and apparent size.[30]

A diagram to illustrate this description of the outgoing visual cone (Fig. 3.5) appears
in the Latin commentary on Cleomedes by Robert Balfour (1605).[31]

The combination of a geometrical approach with a belief in outgoing rays
poses a problem: how do the rays know their own length? Ptolemy argued that
the flux or rays were sensitive, but was unclear on how distance was perceived.
The matter was made clearer in the eleventh century by Ibn al-Haytham, who believed

in incoming rays. He based his own *Optics* (*c.* 1040) partly on Ptolemy's *Optics* and quoted the above passage almost verbatim.[32] He added his own statement on the topic, writing: 'When human vision perceives the size of visible objects, it perceives it from the size of the angles that visible objects project to the centre of vision, and from the degree of intervening space, and by comparing the angles with the intervening space' (our translation).[33] In this passage Ibn al-Haytham clearly describes the rays as incoming: angular size is given by the visual cone, distance is given by other information about the intervening space, and linear size is derived in a geometrical manner.

Descartes (1637b, Sixth Discourse; 1965, p. 107) expressed the same idea even more clearly:

> Their size is estimated according to the knowledge, or the opinion, that we have of their distance, compared with the size of the images that they imprint on the back of the eye; and not absolutely by the size of these images, as is obvious enough from this: while the images may be, for example, one hundred times larger when the objects are quite close to us than when they are ten times further away, they do not make us see the objects as one hundred times larger (in area) than this, but as almost equal in size, at least if their distance does not deceive us.

Since that time size–distance invariance has been widely accepted as the main explanation of size constancy. For example, Desaguiliers (1736a, 1736b) demonstrated size constancy experimentally before the Royal Society of London, and showed that it varied with the distance cues available and with the observer's hypothesis about the distance. His diagram to illustrate size–distance invariance is shown in Fig. 3.6.

The compelling nature of size constancy for familiar objects was commented on by Euler (1762, Vol. 2, p. 484):

> It is not astonishing that our judgement respecting the magnitude of objects should not always be in correspondence with the visual angle under which we see it: of this, daily experience furnishes sufficient proof. A cat, for example, appears, when very near, under a greater angle than an ox at the distance of 100 paces. I could never, at the same time, imagine the cat to be larger than the ox.

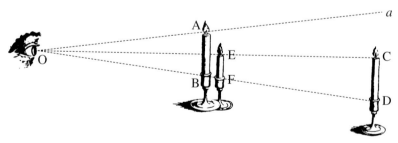

Figure 3.6 Desaguilier's diagram (1736a, Fig. 5) for size–distance invariance. The candle CD is at twice the distance of the candle AB, and subtends half the angle at the observer's eye, but appears the same linear size as AB. The small candle EF appears the same linear size as AB if the observer supposes it to be at the same distance as CD.

The existence of some degree of size constancy has often been confirmed.[34] However, its existence is not sufficient evidence to establish the size–distance invariance hypothesis, since size constancy could be caused by factors other than a change in perceived distance.

Perceptual size–distance invariance

So far we have described size constancy and size–distance invariance in the conventional manner found in most psychology textbooks. However, several authors have suggested that *perceived* angular size should be substituted for true angular size. McCready (1965, 1985, 1986) has formulated this view in detail, pointing out that the classical hypothesis mixes the *physical* measure of angular size with *perceived* measures of distance and linear size. He argues that it is logically preferable to speak of perceived measures for all terms. *Perceptual* size–distance invariance might then be said to hold in all situations. We do not sometimes perceive angular size and sometimes linear size, but, rather, we simultaneously perceive both angular and linear size in addition to distance. A change in perceived linear size may be due to a change in perceived distance alone (as in the classical view shown in Fig. 3.5), or to a change in perceived angular size alone (Fig. 3.7), or more usually to some combination of the two (Fig. 3.8). This contrasts with the classical view of size–distance invariance, where a change in perceived linear size must be due to a change in perceived distance if true angular size is constant.

If perceived angular size is different from true angular size, the question arises as to how it should be measured. Most authors have regarded retinal image size and angular size as interchangeable, as was illustrated in Fig. 3.2: yet the one is linear and the other angular. Many experimenters have used linear size matching techniques, but with angular size instructions, and then calculated the angular size geometrically. If the concepts differ in some important way, this is not a legitimate procedure. There are some more truly angular definitions. One such definition is that perceived angular size is *the*

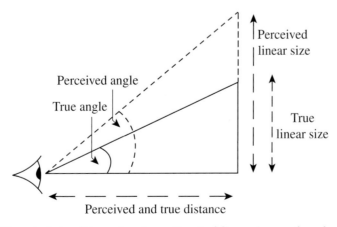

Figure 3.7 Perceptual size–distance invariance. Perceived linear size may be enlarged because perceived angular size is enlarged, while distance is correctly perceived.

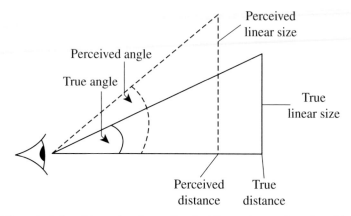

Figure 3.8 Perceptual size–distance invariance. Perceived linear size may be enlarged because perceived angular size is enlarged, while perceived distance is too close.

proportion of the visual field an object appears to fill:[35] observers could be asked to estimate this proportion numerically, or perhaps to make sketches indicating both the visual field and the object.[36] A simpler definition is that perceived angular size relates to the *perceived angle subtended by the object at the eye*, and should be estimated in degrees or matched with an angle shown on a protractor.[37] Other authors, particularly McCready, have argued that perceived angular size has a directional quality: it relates to *the perceived position or angular direction of the outer edges of the object.* For large angles, the direction can be measured by pointing the finger, or similar techniques. Small angles, such as the half degree subtended by the moon, cannot be measured in this way. Medium and small angles would have to be measured by non-directional techniques.

Testing size–distance invariance

Size–distance invariance is hard to evaluate, because most investigators have obtained only one measure of perceived size, and have often failed to specify whether it was intended to be angular or linear. When judgements of both linear size and distance are obtained, the classical form of the hypothesis does not hold in a strict geometrical manner.[38] A question of relevance to the moon illusion is what happens to perceived size when the perceived distance is indeterminate – as is the case for very large distances. In general, size constancy is not maintained, and all far objects appear equidistant and closer than they really are. Such judgements are sometimes described as reflecting angular size; however, this could mean either that the test method involved judgements of angular size, or that the judgements were of linear size but that the perceived distance remained constant. It is sometimes stated that the perceived size represents a compromise between angular and linear size,[39] with linear size predominating at clearly defined distances and angular size predominating at indeterminate distances. Yet it does not make sense to speak of a compromise between two quite different kinds of perceptual size.[40] Instead, one would have to say that judgements of linear size increase with perceived distance, but fail to increase when the distance is indeterminate. Alternatively, one might say that the type of judgement changes in kind when there is inadequate distance information, just as it changes with the type of instructions given to the observer. It should always be possible for an observer to attempt to make an angular size judgement, if indeed perceived angular size is based on the untransformed retinal image without any need to take distance into account.[41] Linear size judgements, however, depend on distance information, and it is perhaps not surprising if without it observers revert to angular size judgements.

The perceptual formulation of size–distance invariance has not been adequately tested. To do so would require judgements from the same observers of all three variables: perceived angular size, perceived linear size, and perceived distance. Some investigators have, however, attempted to measure perceived angular size at different distances. Most have used the same linear size matching techniques as in classical size-constancy experiments, but with 'angular size', 'retinal size', or 'apparent size' instructions. If all other cues are totally removed, size judgements correspond to true angular size;[42] but if some spatial cues are available, correct angular matches can only be made if the two targets are lined up one in front of the other, or side by side, so that they fall on neighbouring parts of the retina at the same time. If these conditions are not met, the matched angular size grows with viewing distance.

An example of growth with distance can be found in the classic size-constancy experiment of Holway and Boring (1941), even in the condition when the spatial cues were reduced by the use of monocular viewing through a peephole into an almost dark corridor.[43] For this experimental condition there were only two observers, who were described somewhat unclearly as matching 'perceived size'.[44] The instructions for an angular match were quite clear in the outdoor experiments of Gilinsky (1955)[45] and Leibowitz and Harvey (1967 and 1969),[46] all of which involved many observers with full natural viewing at large distances. All these authors used a nearer adjustable target which could be matched in perceived angular size to a more distant target: the adjustable target was set too large[47] and the error increased with the distance of the further target. The results of the three sets of experiments (shown in Fig. 3.9) suggest that the overestimation reaches a factor of 4 to 8 in the distance. The overestimation is naturally greater when the adjustable target is closer: the angular ratio of the matched to the standard target starts at 1.0 at the nearest viewing distance and reaches a factor of 3 to 4 when the viewing distance is increased by a factor of 10. However, the overestimation found by Leibowitz and Harvey increases more rapidly than that of Gilinsky at further distances. A possible reason for the difference is that their outdoor scene – a university mall – provided more size and distance cues than Gilinsky's open field.

A different and truly angular method was used by Higashiyama (1992): his observers made numerical angular estimates, or angular matches with a protractor, when viewing target lengths on the wall of a building from distances of 3–30 m. His results show that judged angular size increases with viewing distance when true angular size is held

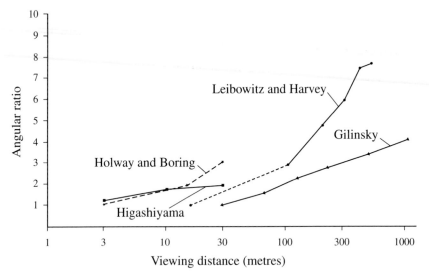

Figure 3.9 The ratio of judged angular sizes of targets as a function of distance. The experiments of Holway and Boring, Leibowitz and Harvey, and Gilinsky show the relative enlargement of a far to a near target, the ratio being 1.0 for targets at the same distance. The experiment of Higashiyama shows the ratio of numerical estimates to true angular size for a 10 degree target.

constant.[48] His data for a 10 degree target are also shown in Fig. 3.9, and they are not affected by the question of the distance of an adjustable target: they lie on a curve that can be extrapolated to a factor of about 3 at 1000 m. A factor of 3 or 4 also corresponds to people's numerical estimates of the extent to which distant scenes appear enlarged in real life compared to photographs.[49]

The fact that perceived angular size grows with distance tells against the classical form of size–distance invariance,[50] but is insufficient evidence on its own to prove or disprove the perceptual form.[51] The perceptual form is preferable in that it allows more sources of error, but many people have difficulty in comprehending it.

There are some alternative approaches to size perception. For example, Bishop Berkeley in 1709 issued a clear challenge to size–distance invariance, objecting to 'the humour of making one see by geometry'. He stated in *A new theory of vision* (Section 53) that we learn to associate certain *cues* with distance, and also with increased perceived size, but denied that size and distance perception were causally linked. Non-geometric theories are not as explicit and testable as size–distance invariance, nor have they given rise to much experimental work on the moon illusion.[52] For the sake of clarity we shall, as far as possible, use the framework of perceptual size–distance invariance to guide our discussion of the illusion. This can cause difficulties when describing the views or experiments of authors who reject a geometrical approach: were they measuring perceived linear size or perceived angular size, or neither? We shall re-examine the suitability of the framework in the final chapter.

Theories of the illusion

There have been numerous attempts to provide an acceptable explanation for 'that most astonishing and perplexing of all natural phenomena' (Gilinsky, 1971, p. 71). In fact the number of theories is embarrassingly large.[53] 'Of all perceptual illusions, the Moon Illusion … has perhaps garnered more descriptions, theoretical analyses, and emotional turmoil than any other' wrote Johannsen (1971, p. 134) in an article on the early history of illusions.

All theories of the illusion are based on the view that it is a normal, rather than a pathological phenomenon: it cannot be explained in terms of either abnormal observers or abnormal perceptual processes. Visual illusions arise when the normal processes of visual perception are inappropriate, or are affected by factors causing unexpected perceptual errors.[54] The different theories of the moon illusion attempt to identify these error-producing factors.

Few serious attempts have been made to classify theories of the moon illusion, though Coren (1989) divides them into those involving either structural mechanisms or strategy mechanisms. We would instead divide them into three classes:

a. The cause lies outside the body of the observer; these physical theories assume the real angular size of the moon to be enlarged, for example by atmospheric refraction. They are discussed in Chapter 5.

b. The cause lies in the eye or other structures of the body; these physiological or structural theories are discussed in Chapters 6, 7, 11, and 12.

c. The cause lies in the brain processes associated with perception; these psychological or neurophysiological theories are discussed in Chapter 6 and Chapters 8 to 12.

This classification should not be too rigidly applied: several theories changed from one type into another, or contain aspects of more than one class. However, it represents a useful framework within which the numerous attempted explanations can be presented.

It is interesting to relate the two geometrical assumptions about perceived size to this framework. The first assumption (that perceived size corresponds to angular size) includes all the physical and physiological theories, as well as those perceptual theories that claim the moon's perceived horizon enlargement is due to angular size contrast or the grain of the visual field (Chapter 10). In these theories the real or the perceived angular size of the moon is considered to be the primary datum on which judgements of its enlargement are based. Its perceived distance usually plays no role: it is at most considered to be an effect, rather than a cause, of the illusion. The horizon moon, seen as angularly enlarged, is predicted to appear closer.

The second assumption (that perceived size corresponds to linear size) is usually taken to imply that the real and the perceived angular size of the moon are not subject to illusion. The moon's perceived size is not determined directly by its angular size, but by the observer's interpretation of this in relation to perceived distance. Different theories of this type ascribe variations in the moon's perceived distance to factors such as aerial perspective (Chapter 6), the flattened appearance of the sky (Chapter 8), or the perceived distance of the horizon (Chapter 9). All of them imply that the horizon moon, perceived as enlarged in linear size, must appear further away. The predictions are, of course, no longer clearcut if we adopt the perceptual version of size–distance invariance, in which both distance and angular size are subject to perceptual illusion.

It is often claimed that the two classical assumptions about perceived size fit neatly into what used to be called the nativist/empiricist debate – the angular size assumption being nativist and the linear size assumption being empiricist. In modern terms, nativist accounts emphasize *data-driven* or *bottom-up* factors: they portray perception as determined mainly by sensory input rather than by experience. Empiricist accounts emphasize *conceptually driven* or *top-down* factors: they argue that sensory information is interpreted by knowledge and expectations based on previous experience. However, there is no logical reason why assumptions about angular or linear size should correspond to this division: it is quite possible that experiential factors could affect the perception of angular size, while the scaling of perceived linear size by perceived distance could be innate.[55] Both bottom-up and top-down processes contribute to most mental activities, and presumably also to our perception of the moon. We will attempt to unravel some of these factors in the course of this book.

Summary

Subjective measures of size are defined: the perceived size of an object may be measured by numerical estimates, or by matching it to some other object of known size. Perceived size may be analogous to linear or to angular size, or to neither. The perceived size of the sun and moon is often described as about one foot in diameter, though the meaning of this is obscure. The Epicureans believed that the sun was the size that it appeared to be, but this idea was challenged on various counts. Some authors argued that perceived size was determined by relative size, and others argued for a geometrical approach. The classical form of size–distance invariance states that perceived linear size is determined by true angular size and perceived distance; on this theory the perceived size of the sun should vary with its perceived distance. Size–distance invariance first appears in the writings of Ptolemy (*c.* AD 170) and Cleomedes (*c.* third century), and was developed by Ibn al-Haytham (*c.* 1040). The perceptual form of size–distance invariance was developed by McCready (1965), and replaces true angular size by perceived angular size. Neither formulation is supported by experiments, and measurements show that perceived angular size normally grows with viewing distance. Geometrical models of perception were denied by Berkeley (1709), who argued that size and distance perception were learned independently. Theories of the moon illusion can be classified as physical, physiological or perceptual. Physical and physiological theories imply that there is a real angular enlargement of the horizon moon. Perceptual theories imply that there is a perceived angular enlargement, or a perceived distance enlargement, or some combination of these.

CHAPTER 4

Measuring the moon illusion

In the last chapter we defined various meanings of size and described how they relate to different theories of size perception. These theoretical differences are important, because they affect the way in which experimenters attempt to measure and explain the moon illusion: method and theory are inextricably linked. Even the early naturalistic observations were tied to some theory about perception, but the link became tighter with the growth of systematic experiments. We will first describe the earliest attempts at experimental observations, and then consider what can be concluded from drawings of the sun and moon.

The first experiments

Several attempts have been made to measure the perceived size of the sun or moon at different elevations, in order to calculate the magnitude of the celestial illusion. The first attempt seems to have been made by Eugen Reimann (1902b), who used himself and a colleague as observers. The experiment was carried out on the beach, where the sun could be observed both high in the sky and while setting over the sea. Reimann argued that if he could obtain some tangible representation of the perceived size of the sun around noon, and again at sunset, then these could be used to measure the sun illusion directly. He attached a white paper disc with a diameter of 340 mm to a pole so that it could be viewed at eye level. The observer looked at the sun through a dark glass to protect his eyes and formed an impression of its size. He then shifted his gaze to the paper disc, which was usually placed in the opposite direction to the sun, and approached it or walked away from it until it appeared to have the same size as the sun. His distance from the disc was then measured. To ensure that the results would not be influenced by the starting point, half of the judgements were made by starting quite close to the disc and then retreating from it, and the other half by starting a considerable distance away and approaching it. Measurements were made on several days, at noon when the elevation of the sun was at its maximum of about 55 degrees, and just before sunset.

The magnitude of the sun illusion, or the ratio of the matched sizes of the sun at sunset and at noon, was obtained by calculating the ratio of the measured distances at noon and at sunset. The results were 3.1 for Reimann himself and 3.6 for the other observer. The angular size of the disc when judged equal to the noon sun was on average 30.7 minutes of arc, as calculated from the noon distances to the comparison disc. This is very close to the real angular size of the noon sun at the time, namely 31.7 minutes of arc. Thus the perceived size of the sun at high angles of elevation was found to correspond to its real angular size, while at sunset it was estimated to be enlarged more than three times.

However, it is not immediately obvious what Reimann succeeded in measuring. First we note that the comparison disc had a fixed real linear size. The observers could, if they had so wished, have estimated the disc's linear size quite accurately at all distances, since our perceptual processes compensate effectively for varying distance in the range of 10 to 40 m used here. However, the perceived size of the sun that Reimann measured did not behave like a linear size, because it was very different at noon and sunset but was judged equal to the same disc (at different distances). In fact the method and results indicate that the sun's judged linear size was either the same (namely 340 mm) at noon and sunset, or was not attended to. Instead, the perceived size that Reimann defined by his method of measurement was clearly analogous to angular size.

When stated in the above manner, this conclusion may appear to be both obvious and trivial, but in fact it is neither. Despite the analogy with angular size, Reimann and several later investigators interpreted these and similar results as proving an enlargement of the perceived linear size of the sun. We will return to this confusion in later chapters.

A second point worth noting is a comment by Reimann on the nature of the comparison process which, he says, was obviously not intended to compare the real angular size of the sun to that of a disc seen in the same direction and at the same time. The disc was therefore placed behind the observer or to his side when facing the sun, and the comparison was supposed to be between the two sense impressions: 'while we devoted ourselves as unbiased as possible to the sensory impression [of the sun], we distanced ourselves [from the disc] until it appeared to have the same size as the sun' (Reimann 1902b, p. 162).

Reimann's results can be summarized as follows: the matched angular size of the sun was close to its real angular size at high angles of elevation, but was enlarged more than three times near the horizon. Similar measurements were attempted by other investigators. Pozdena (1909) compared the full moon to circular patches of light, carefully made to match the moon in colour and brightness. A series of these circles was used, ranging in size from 10 to 100 mm, presented at a fixed distance of 4 m. The observer had to select one which matched the moon in perceived size. Again the description of the procedure clearly implies the matching of perceived angular sizes. With the moon at its highest elevation of about 30 degrees the matching circles chosen by three observers had angular diameters of about 35 minutes of arc on average – slightly larger than the real angular size of the moon. The horizon enlargement factors were found to be 2.4, 2.5, and 2.6 for the three observers.

The same problem of interpretation arises as for Reimann's results.[1] In his procedure Pozdena ignored the immense difference in distance between the comparison discs and the moon, thus defining perceived size as judged angular size. Nevertheless, he ascribed the horizon enlargement to changes in the moon's perceived distance – changes which are surely very much smaller than those he had just ignored when making his measurements.

Less consistent results for the size of the illusion have been found by others using the same basic technique. Dember and Uibe (1918) employed the method at Tenerife[2] in the summer, when the maximum angle of elevation of the sun was about 84 degrees. They found the illusion to vary from day to day. Stroobant (1928) compared the full moon to a disc of adjustable size placed 38 m from the observers. The latter were all astronomers, who often had to estimate angular sizes and distances on the celestial sphere during the course of their daily work. These observers matched the moon to a disc of about the same angular size at both low and high angles of elevation, thus providing estimates of real angular size rather than perceived size.

In contrast, Holway and Boring (1940a) found the angular size of the full moon to be overestimated at both high and low angles of elevation, though more so in the latter case. They matched the perceived size of the moon to that of an adjustable circle of light projected onto a screen 3.5 m from the observer. Three observers, including the authors, made matches on two days, with the moon at elevations of less than 5 degrees and at its maximum of between 50 and 60 degrees. The resulting angular sizes of the comparison circles varied between 4 and 6 degrees at the lowest elevations, and between 2 and 5 degrees at the highest elevations, giving a mean relative enlargement for the low moon of 1.7 (with individual enlargements ranging from 1.2 to 2.0).

As a last example of this type of measurement and the problems associated with it we note a more recent experiment by Solhkhah and Orbach (1969). They asked their observers to match the perceived size of the moon to an adjustable disc:

> The comparison disc was exposed beside the moon at horizon on one occasion and zenith on another at distances of 10 to 300 ft from the observer. It quickly became apparent that our subjects were making judgments based on visual angle and we had to conclude that our method was unsuccessful in measuring the magnitude of the moon illusion. The visual angle approached 30 minutes of arc regardless of the position of the moon in the sky. Yet on further interrogation each of our subjects agreed that the horizon moon looked several times bigger than the zenith moon (pp. 88–9).

Clearly Solhkhah and Orbach obtained accurate matches to the real angular size of the moon, just as Stroobant had. In Stroobant's case one of the reasons may have been the observers' ingrained habit of estimating angular size, while in this later experiment the proximity of the comparison disc and the moon made it impossible for the observers to attend to anything except their real angular sizes – an error that Reimann and other earlier investigators carefully tried to avoid.

Criticism of these techniques

What can be learned from these attempts to measure the celestial illusion? First, the measurement methods contained a very large difference in viewing distance between the moon and the comparison discs. In choosing to ignore this difference, the investigators assumed that the perceived size of the sun or moon was adequately represented by its judged angular size, and that perceived distance (over this large range) played no role at all in determining perceived size. Yet, as we noted in the last chapter, many people express their perceived size of the sun and moon in terms of length measures. The above experiments can therefore be criticized for concentrating on the angular aspect of the illusion while excluding the linear aspect.

Second, with this type of measurement the perceived angular size may correspond to the real angular size of the moon or may be enlarged. When the comparison disc is seen right next to the moon, real angular sizes are matched and no illusion is shown. When the comparison disc and moon are not seen at the same time, the illusion is found to be present and may reach a factor of up to 3 or more. The size of the illusion is therefore affected by the closeness of the moon to the matching disc, and whether the observer has to rely on memory for another occasion. This may help to explain why the illusion is generally small when comparing the separations between pairs of stars at different elevations; the comparison is usually fairly direct, as the two arcs can be viewed in quick succession or even simultaneously (Chapter 1).

There is a further measurement difficulty that affects these and many other experiments. It is known as the *error of the standard*, and we mentioned it in Chapter 3 (note 47) when discussing the measurement of size constancy. If one target is the standard (fixed in size), and other variable targets are matched to it, there may be a tendency to overestimate the standard and thus set the variable too large. In the experiments we have just described, the moon or sun was always the standard, and some other target was matched to it. This difficulty is unavoidable if the real moon or sun is used in an experiment; but with artificial moons it should be possible to counterbalance the test procedure, so that both targets are varied in size.

A moon machine

In the late 1950s, Lloyd Kaufman and Irvin Rock concluded that adjusting the size of a nearby disc to match that of the 'infinitely' distant moon may cause problems, for some of the reasons just stated. The authors[3] developed an elegant solution to the distance problem by allowing observers to adjust the size of an artificial moon at one place in the sky to match that of another at a different place in the sky. An artificial moon was created by placing a luminous disc of variable size at the focus of a lens. The lens refracts the light from the disc so that, when one looks into the lens, the light appears to be coming from a disc at an infinite distance. To avoid the need to look directly into the lens, a thin plate of glass was mounted at an angle of 45 degrees to the optical axis of the lens: this enabled observers to view the reflected image of the disc through the glass. This 'virtual moon' was perceived as being at the distance of objects on the horizon, and similarly at the same distance as the real moon when elevated in the sky. In fact, to see the virtual moon clearly, the observer's eyes had to be focused at infinity, as they must for any very distant object. Figure 4.1 is a schematic diagram of the original setup employed by Kaufman and Rock. It shows two optical systems, one allowing the subject to view a moon over the horizon, and the other allowing him to view a moon at a higher elevation.

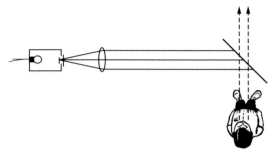

(a) Arrangement for viewing horizon moon

(b) Arrangement for viewing zenith moon

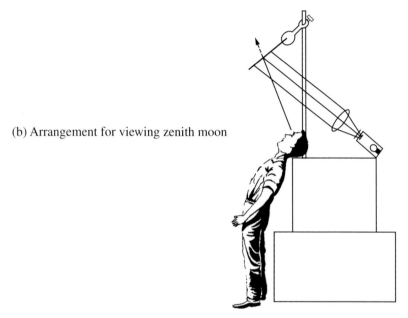

Figure 4.1 The first apparatus devised by Kaufman and Rock for projecting artificial moons in the sky. (From Kaufman 1974, p. 362, with permission of the author and publisher.)

As their work progressed, Kaufman and Rock took their moons to different locations so that they could explore the effects of factors such as different terrains and different distances to the visible horizon. They found it necessary to develop a more portable version of the apparatus. The original drawing of this portable apparatus is reproduced in Fig. 4.2. While the lenses of the first apparatus were large enough for both eyes to see the reflected images of virtual moons in the combining glass, in this apparatus the lens was smaller and only one eye could see the moon. However, with both eyes open observers saw the terrain and sky with two eyes, and were unaware that the moon was imaged in only one eye. Under comparable conditions, the two types of moon machine gave essentially the same results.

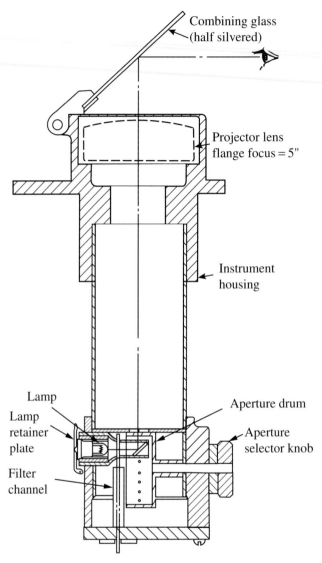

Figure 4.2 The portable apparatus developed by Kaufman and Rock. (From Kaufman 1960, with permission of the author.)

In the course of their work the authors found that it was important to counterbalance which moon is used as the standard and which as the variable: they calculated that the variable was adjusted larger than the standard by about 11 per cent – quite a large amount in relation to some of the effects that may contribute to the moon illusion.

With this technique the problem of the different perceived distances of the moon and comparison discs is removed. Both artificial moons are at an infinite distance, so the magnitude of the illusion will not be affected by whether the observer adopts an angular or linear attitude to perceived size. However, some measurement problems still remain.

One possible problem is that, when two artificial moons are visible at the same time, the illusion could be smaller than when the real moon is seen in isolation and compared with the memory of another occasion. This should be investigated. Also, as discussed in earlier sections, if the comparison disc is not too far above the horizon disc, the illusion may disappear because the observers match the real angular sizes of the two stimuli. Other problems concern the different biases that arise with different methods of adjusting the size of the discs.[4]

Several types of moon machine have been used in various visual scenes by different investigators, so it is hard to say how far their machines are responsible for their differing results. The horizon enlargement found by Kaufman and Rock (1962a) for ten observers varied from zero to a factor of 2.03, with a mean of about 1.5; and the mean enlargement under various distances and cloud conditions fell in the range 1.28–1.58 (Rock and Kaufman 1962). Hamilton (1965, 1966) used a similar instrument, and obtained mean illusion ratios with a maximum of only 1.36; but his observers were viewing from a height, which reduces the illusion (Chapter 12). Iavecchia, Iavecchia, and Roscoe (1983) used a different technique (Chapter 7) and found a maximum of only 1.26. These values are much less than the factor of 2 or more given as numerical estimates for the natural moon illusion by earlier authors (Chapter 1). However, we cannot be sure how far numerical estimates represent what is actually perceived, since such estimates are also subject to various biases. Values of around 1.5, as found by Kaufman and Rock, are similar to the results of several other experiments yet to be described.

It has been suggested that a different technique may reproduce the large numerical estimates of the natural moon illusion: namely, drawings showing the sun or moon in a natural landscape or other measurable context. We turn now to a consideration of the celestial bodies in art.

The sun and moon in art

The sun and moon are often depicted as extraordinarily large, particularly when low on the horizon. However, we cannot interpret the drawn size as a representation of perceived size without making some assumptions about the meaning of size in pictures.[5] We would have to assume that the artist was intending to represent perceived size, and was doing so according to some rules of projection that specify linear or angular size.

The first assumption is often untrue. Size may denote the social importance of the chief subject of the picture. For example, the donors of a religious painting are shown as small relative to Christ, and effigies of knights in armour may be accompanied by small wives and miniature dogs. Size may also denote emotional significance or interest. For example, experimental studies have shown that children draw Santa Claus and Christmas trees larger before than after Christmas.[6] So an enormous sun or moon may represent its emotional or religious significance rather than its true or perceived size. Such pictures are not useful for measuring the moon illusion.

The second assumption needs further clarification. Firstly, the artist may be attempting to represent true linear size or true angular size rather than any kind of perceived size. Secondly, he may be using one of several projection systems, or perhaps a hybrid system. The techniques of western artists have been heavily influenced by the rules of perspective, but there are several other projection systems in use. Willats (1997) has clarified these by using engineering and computing terms. He distinguishes three main projection systems: *oblique* (or *pictorial*), *orthographic* (or *orthogonal*), and *perspective*. In oblique projection, orthogonal lines are parallel and run at an oblique angle, and size is unchanged with distance; this system is common in oriental art. In orthographic projection there are no orthogonal lines, and size is unchanged with distance; edges perpendicular to the picture plane are shown as points, and planes perpendicular to the picture plane are shown as lines; this system is used for engineering drawings, and also appears in Greek vase paintings, friezes, and some other types of art. In the perspective system, orthogonal lines converge to a vanishing point, and size reduces proportionately for distant objects; this system is common in Renaissance art. To complicate matters there are many variant and hybrid systems. There are also some unusual systems, such as *inverted* or *reversed* perspective, in which the orthogonal lines diverge with distance, and sizes become larger; this may occur in some Cubist paintings, and in Byzantine art and Russian icons.

We illustrate below (Fig. 4.3) what the three main systems would imply for the represented size of the sun or moon. In oblique projection (a), and in orthographic

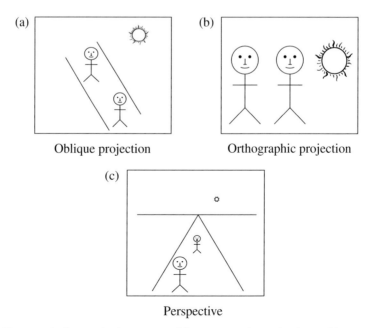

(a) Oblique projection

(b) Orthographic projection

(c) Perspective

Figure 4.3 The moon in three projection systems: oblique, perspective, and orthographic.

projection (b), size is unchanged with distance, so the moon is represented as having its perceived linear size, which is about the same size as a man's head. In the perspective system (c), the moon is represented in proportion to its true angular size, and is therefore smaller than the head of a man in the foreground. Pictures drawn in this system are clearly the most useful for measuring the moon illusion.

The development of perspective

Many books have been written on the way in which the rules of perspective were outlined in classical times and refined in the Renaissance when Europeans rediscovered the earlier literature.[7] Linear perspective was implied by Euclid[8] who stated in his *Optics* (6th Proposition) that parallel lines come closer together in the distance: 'Parallel lines, when seen from a distance, appear not to be equally distant from each other.'[9] We know that architects in classical times were aware of linear perspective, since Vitruvius (first century BC) stated that the slope of the upper parts of columns must be altered to counteract perspective effects. Perspective was also used in an approximate manner by artists in the Middle Ages. However, refined mathematical techniques were not invented until Renaissance times. Leon Battista Alberti wrote a book in 1436 entitled *Della pittura*, in which he described how linear perspective could be constructed. He suggested that a scene should be viewed through a window frame and the outline of the objects traced on the pane, and this came to be known as 'Alberti's Window'. Artists from the sixteenth to nineteenth centuries, such as Caravaggio, Velazquez, and Ingres may have used lenses and mirrors to make accurate portraits. Seventeenth-century artists such as Vermeer may have used a camera obscura to obtain a true perspective image of the scene they wished to paint.[10] The advent of photography in the nineteenth century made such techniques unnecessary.

Early art

The earliest known human art is found in cave drawings in France dating from the Ice Age. The compositions generally use oblique or orthogonal projection, and distance is sometimes indicated by overlapping objects. For example, there are drawings from about 23 000 BC at the Peche-Merle cave showing animals superimposed, and given scale by the sprayed outline of a real human hand.

The earliest representation of the sun that we have come across occurs in an Ice Age inscription on a bone found in a cave at La Vache in France: it consists of a line drawing of human figures and animals, with a circle that may represent the sun or moon (Marshack 1975). This circle is about the size of a human head (Fig. 4.4), which agrees with the ancient observation that the sun is the size of a foot. The drawing does not show linear perspective, so we would have to take it as a representation of perceived linear size.

Representations of the perceived linear size of the sun or moon can be found in many types of art, but they are difficult to interpret in perceptual terms for the reasons that we

Figure 4.4 Ice Age sketch on a bone from La Vache in the Pyrenees, showing human figures and the sun or moon. (Redrawn from Marshack 1975.)

discussed in Chapter 3. They are not of much help in measuring the moon illusion. We therefore turn to the more promising field of perspective.

The analysis of moon perspectives

Perspective representations of the sun and moon may be used to measure the size of the celestial illusion, if the depicted size is greater than the true perspective size. Careful measurements are necessary in such pictures, because the pictorial distance cues – such as converging lines – produce a small perceptual enlargement even when the depicted size is not enlarged.[11] Tolansky (1964, pp. 101–18) selected pictures for which it was possible to estimate the distance between the artist's viewpoint and objects of known size in the foreground. In that way the angular size of these objects can be calculated, the angular scale of the picture determined, and the angular size of the depicted moon measured. Paintings of the interiors of houses, or of buildings across a street, are useful in this respect. Landscapes are less informative, but if the moon and foreground are included one can calculate whether the scale makes sense for the whole scene. If only a portion of the distant landscape has been drawn large, as in a photograph taken with a telephoto lens, the sun may be represented as equal in size to a large object such as a house. That can occur if the image size of the house is half a degree in the distance, the same as the moon: the relative size of house and moon looks impossible when no foreground is included, and the house is taken to be near (Chapter 5). A telephoto picture of the moon in relation to standing stones (Fig. 1.1) illustrates the enlarging effect.

Tolansky analysed several pictures by famous nineteenth-century artists. He examined the drawing *The Bluestocking* by the French artist and caricaturist H. Daumier. This drawing shows a woman sitting in front of a window, with a large moon visible through the window just above some buildings. The diameter of the moon is enlarged by a factor of 4 relative to the probable perspective from the artist's viewpoint. He also examined the painting *Coming from Evening Church* by the English painter Samuel Palmer. This painting shows a church with a procession of people in the foreground, and a huge moon just above a hill in the background. The scaling of the relative parts of the picture is impossible, unless the moon's diameter has been exaggerated by a factor of 5. The painting *The Sower*, by the Dutch artist Van Gogh, shows an enormous setting sun (or possibly rising moon) on a flat horizon, with the sower in the foreground and a farmhouse in the distance. The sun or moon has been exaggerated by a factor of more than 10. Tolansky suggests that these artists exaggerated the moon to make it look more impressive, and not because they really saw it that way. He also examined the painting *The Carnival Evening* by the French artist H. Rousseau. This picture shows a carnival couple in front of a large house and tall wood, with a high full moon at the top of the picture. Tolansky argues that the moon is realistic in size, because Rousseau prided himself on his realism. Actually, the moon measures 0.625 degrees on the scale calculated by Tolansky, showing an enlargement by a factor of 1.25. This may well be a faithful representation of the perceived size of the fairly high moon, seen above the highest trees. It is not really possible to calculate the true elevation of this moon, because the vertical scale of the picture seems to have been elongated in a manner reminiscent of El Greco.

The pictures described above show varying degrees of moon enlargement. That may reflect the varying temperaments of the artists, or it may reflect the way the celestial illusion decreases with the elevation of the moon. Tolansky thought that artistic licence was the more important factor. However, if we wish to use drawings to measure the moon illusion, we need to select artists who are attempting to be realistic, and we need to know the angle of elevation of the moon.

Tolansky (1964, p. 116) argued that the perspective scale of everything in these pictures was correct, apart from the exaggerated size of the moon. He wrote:

> the curious feature about the whole situation is that the exaggeration applies to the moon and to the moon only. The rest of the picture is consistent; all the terrestrial objects are correct in relation to each other. Sun, moon and constellations of stars are the only objects not on the earth which deviate because they are the only objects which subtend a uniform angle as they approach the horizon.

We doubted that it was true that realistic artists normally draw terrestrial scenes in correct perspective, and therefore analysed a sketch by a twentieth-century artist at an identifiable and accessible location. We selected a sketch of the village of Temple, 20 km south of Edinburgh (a latitude of 56 degrees north), drawn by the Scottish artist William Gillies in about 1967. The viewpoint was just outside the artist's house in the main street, looking south-south-east (azimuth 160 degrees). The season was winter or early spring (since there are no leaves on the trees in the picture), and the sun is depicted

(a)

(b)

Figure 4.5 (a) Drawing of the sun in *Village Street, Temple* by William Gillies *c.* 1967. It is a winter scene, looking south, probably sketched on a February morning. (Reproduced by permission of the Royal Scottish Academy, Edinburgh.) (b) Photograph of the same scene in 2000. Taken with a digital camera with a 5.4 mm lens (equivalent to a 35 mm lens in a conventional 35 mm camera). The moon was photographed similarly, and the image superimposed to the same scale in the same position as in the drawing. (Photographs by Helen Ross. Computer combination by Peter Hancock.)

at the height of a house across the road (Fig. 4.5a), which we measured as an altitude of 18 degrees. We calculate that the sun had a declination of minus 14.1 degrees (i.e. 14.1 degrees south of the celestial equator), and that the sketch was therefore made on 11 February at about 10.42 a.m. The scene has changed little in 30 years. A recent photograph from the same spot is shown for comparison (Fig. 4.5b). The photograph was taken with the widest possible angle on a digital camera, which was equivalent to a 35 mm lens on a 35 mm camera (magnification = 1). It was not possible to reproduce Gillies' perspective, which is like that of a very wide angle lens for the foreground and a telephoto lens for the background. Gillies has also shrunk the relative size of the houses on the right-hand side of the road. We have scaled the photograph so that the large house on the left is the same size as in the sketch. We have also inserted a full moon in the same location as Gillies' sun, with the moon photographed on the same scale. The sun in the drawing is enlarged by a factor of 2 compared to the photograph. The trees in the far distance to the left of the road are also enlarged in height by a factor of 2. Gillies' distant cypress tree on the right-hand side of the road is not enlarged relative to the photograph, but that type of tree will have grown much taller in recent years.

Gillies was probably attempting an approximate perspective drawing, since he includes scaled sketch lines. He enlarges the sun to the same extent as other objects in the distant part of the picture: that is, a factor of 2 relative to the mid-ground, and up to 3 relative to the immediate foreground. Without the use of optical or mathematical devices artists rarely draw scenes in correct perspective, probably because they are influenced by perceptual size constancy.[12] We can therefore discount Tolansky's idea that nineteenth-century artists normally drew terrestrial scenes in correct perspective, but used a different perspective for the celestial bodies.

Modern artists often experiment with various forms of perspective. One of these is aerial perspective – changes in contrast and colour with distance – which we discuss in more detail in Chapter 6. Aerial perspective can be used to enhance or contradict linear perspective. Willats (1997, p. 235) makes the point that contradiction will flatten a picture. He describes Paul Nash's painting *Pillar and Moon* (*c.* 1932), in which the perspective effect of a receding line of trees is counteracted by the warm pink of the distant clouds, and by the bright white of the moon. A similar painting by Paul Nash, *Landscape of the Vernal Equinox*, shows a raised pale white moon and a low red sun of the same size, but the low sun appears larger and nearer despite the foreground depth effect of the converging lines of trees.

The drawings of Cornish

Most artists have no intentions of providing data for the measurement of the moon illusion. An exception to this was Cornish, who was very interested in the perception of the landscape. He spent much of his time drawing sketches of the landscape, and he measured the sketches to see how his so-called 'field of attention' varied in size. He thought (1935) that this variation could explain many phenomena of outdoor perception, and he later (1937) extended this to the sun illusion. He claimed that a restriction of the field of attention leads to an apparent enlargement of the objects within it – an idea similar to the effects of a telephoto lens. For flat or distant horizons, the size of the arc included in sketches was relatively small (Fig. 4.6a); while for nearby mountainous horizons the arc was larger (Fig. 4.6b). Cornish thought that the reason for this was that mountains caused the eye to take in a greater vertical extent: 'It seems, therefore, that

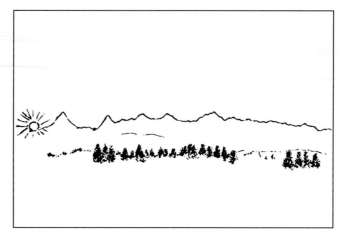

(a) The rising sun from Bern

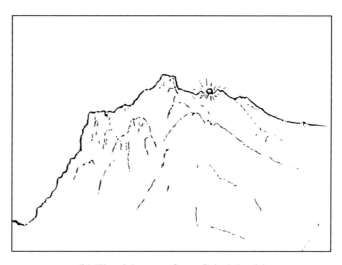

(b) The rising sun from Grindelwald

Figure 4.6 (a) Cornish's sketch of the sun from Bern, rising above the Bernese Alps 40 miles away. The arc of the horizon corresponds to 21 degrees, and the sun in the sketch measured 0.16 inches on a 7-inch page. (b) Cornish's sketch of the sun from Grindelwald, rising above the Wetterhorn four miles away. The arc of the horizon corresponds to 34 degrees, and the sun in the sketch measured 0.09 inches on a 7-inch page. (Reproduced from Cornish 1937, with permission from *Nature.*)

the more the eye takes in vertically the more it takes in horizontally and the less imposing are both dimensions.' But for a long line of low hills he found that: 'whereas the impression of lateral extent received in [this] case was the greater, the arc of the horizon comprised in the panoramic sketch was much less, that is, the eye was sooner satisfied when it had to deal mainly with one dimension.' He concluded from measurement of the sketches that the perceived enlargement of the horizon sun was proportional to the reduction in the true angular size of the terrain included in the sketch. He calculated that the ratio of the diameters of the suns in the two sketches was 1.7, and that of the true horizon arcs 1.6. These ratios are very similar.

We discuss the idea of a 'field of attention' in more detail in Chapter 10. Meanwhile it is worth noting that the enlargement calculated by Cornish from his drawings is smaller than the factor of 2 or more found in the Gillies drawing. The method of calculation is of course quite different: Cornish compared the depicted extent of two different horizon scenes, while in the Gillies drawing we compared the angular size of the foreground and background. The Gillies drawing lacks the presence of a low sun and a high sun; so, while it gives some measure of size constancy, it does not provide a measure of the celestial illusion. A more direct measure of the illusion was obtained by King and Hayes (1966), who measured drawings of the sun on the horizon and in the meridian by 347 student observers: they found variable results, but a mean horizon enlargement of about 1.5. This value is quite similar to that found by Kaufman and Rock with their moon machine.

We have argued in this chapter that the method of measuring the perceived size of the real moon by matching it to the perceived (angular) size of a nearby target gives results that are hard to interpret; data for small numbers of observers show enlargements varying from zero to 3.6 for the low moon compared to the raised moon. More consistent results are obtained by comparing moon drawings or artificial moons low and high in the sky: these give an average size of the celestial illusion of about 1.5. We turn in the next chapter to one of the oldest explanations for that effect: namely, that the real angular size is enlarged by atmospheric refraction.

Summary

Early measurements of the moon illusion involved matching the perceived angular size of a nearby target to that of the moon, though the investigators sometimes claimed to measure perceived linear size. These measurements usually show a small angular enlargement of the raised moon, but a relative angular enlargement of the horizon moon by a factor of up to 3 or more. Moon machines have been used to project images of artificial moons onto the sky at high and low elevations, and adjust one to match the other: they show average horizon enlargements of about 1.3 to 1.5. Drawings of the moon can provide measurements of the illusion, if the artists are attempting to represent perceived size. Early artists used oblique or orthographic projection systems, in which size does not change with distance: they often depicted the sun or moon as about the size of a human head, which represents its perceived linear size. Later artists used linear perspective, in which sizes are proportional to angular size. Some non-realistic artists enlarge the celestial bodies near the horizon by a factor of 10 or more. Artists who attempt to draw perceived perspective usually increase the relative size of the background scene by a factor of 2 or more compared to the foreground, and enlarge the low moon in the same proportion. Sketches also show an enlargement of the horizon sun relative to the raised sun of about 1.5. The size of the illusion thus varies with the method of measurement, but typical measured values are closer to 1.5 than to the factor of 2 or more given in earlier numerical estimates of the natural moon illusion.

CHAPTER 5

Atmospheric refraction

We established in Chapter 2 that the horizon enlargement of the celestial bodies cannot be explained by variations in their real distance. Astronomers have known this since at least the time of Hipparchus in the second century BC, and they have never seriously attempted such an explanation. Many scientists nevertheless thought the enlargement was real, and they have offered only one type of physical explanation – namely, that it is due to an enlarging effect of the earth's atmosphere. This idea developed from a rudimentary beginning during the fourth century BC to a fairly explicit theory some 500 years later, and then gradually degenerated to its more primitive form over many centuries. Yet so compelling is the illusion that even in modern times many people regard the enlargement as real, and attribute it to some atmospheric effect.[1] The same point was made again more recently by Walker (1978a) after studying the opinions expressed by readers of the technical journal *Optical Spectra*. The history and eventual fate of this theory are therefore well worth reviewing.

Atmospheric vapours

Aristotle, in his book on celestial phenomena written during the fourth century BC, provides the first known account of the atmospheric theory. In it he refers to 'stars', and it is worth noting that this term includes the moon and planets, and perhaps the constellations too. When writers claim that stars on the horizon are enlarged they almost invariably mean the moon, sun, and constellations and not individual fixed stars. Aristotle's discussion of the enlargement is quite brief:

> Distant and dense air does of course normally act as a mirror in this way, which is why when there is an east wind promontories on the sea appear to be elevated above it and everything appears to be abnormally large; the same is true of objects seen in a mist, or twilight – for instance the sun and stars which at their rising and setting appear larger than at their meridian.[2]

It is not entirely clear what Aristotle had in mind, but his explanations of other phenomena before and after this passage are of some help. In this particular book Aristotle subscribes to the theory that vision is caused by visual rays leaving the eye through the pupil and returning to the eye after encountering an object, thus bringing an image back to the observer. This idea was an important part of several ancient theories of vision, two of its principal supporters being the mathematician Euclid (*c.* 300 BC) and Ptolemy

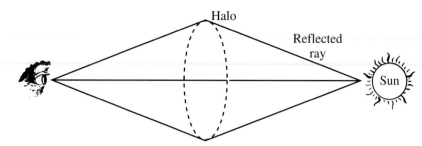

Figure 5.1 Aristotle's explanation of a halo round the sun. Some visual rays are reflected by mist in the atmosphere on their way to and from the sun, forming a second image of the sun in the form of a halo round the primary image.

(second century AD). However, the idea is not usually associated with Aristotle: indeed, in one of his other books[3] he argues against it, calling it 'wholly absurd'.

With this theory as background, Aristotle explains the presence of a halo round the sun in misty weather by arguing that some of the visual rays from the eye are reflected by the mist before they reach the sun, thus reaching the sun by an indirect route. Returning along the same path, they cause us to see a second image all round the directly perceived sun (Fig. 5.1). He seems to have had the same mechanism in mind for explaining the sun illusion, although in this case only one (enlarged) image is seen, so that the reflected image must be taken to mask the directly seen sun completely. The interpretation of this part of the theory is, however, uncertain.

Aristotle's theory implies that the angular enlargement of the setting sun is real, and is caused by processes that are perhaps similar to what we would now call reflection in an enlarging mirror, or to the refraction of light when it passes from one medium to another of different optical density. Refraction is the way that most later authors have interpreted Aristotle's description.

The enlargement was mentioned by Posidonius (*c.* 135–50 BC), an influential Greek philosopher, historian, and scientist, whose works survive only in fragments.[4] According to the Greek geographer and historian Strabo (*c.* 63 BC – AD 25), Posidonius said that the enlargement was due to the vapour in the air, which caused the visual rays to be broken and the sun to be magnified. Strabo wrote:

> Posidonius says that the appearance of the sun's size increases to the same extent at sunset and sunrise over the oceans because of the greater quantity of vapours rising from the waters. Sight is broken through these vapours, as through a [magnifying] glass, and the appearance is enlarged. It is just like when the setting or rising sun or moon is viewed through a dry, thin cloud, when the celestial body also appears slightly red.[5]

The analogies with reflection in mirrors and refraction in water were also repeated by Lucius Annaeus Seneca (*c.* 3 BC – AD 65), a Roman statesman and philosopher. He wrote:

> I have already said there are mirrors which increase every object they reflect. I will add that everything is much larger when you look at it through water. Letters ... are seen larger and

clearer through a glass bowl filled with water ... Stars appear larger when you see them through a cloud because our vision grows dim in the moisture and is unable to apprehend accurately what it wants to ... Anything seen through moisture is far larger than in reality. Why is it so remarkable that the image of the sun is reflected larger when it is seen in a moist cloud, especially since this results from two causes? In a cloud there is something like glass which is able to transmit light; there is also something like water.[6]

Atmospheric refraction

The explanation of the effects of vapours gradually changed from a vague analogy with mirrors to a more recognizable account of refraction. This type of explanation for the celestial illusion was mentioned by Ptolemy in his influential book on astronomy written about AD 142. This book came to be known as *The Almagest* by Arab scientists and, as a result, by Europeans too, although its original title meant 'Mathematical composition'. Ptolemy explains the theory as follows:

> It is true that their sizes [of the celestial bodies] appear greater at the horizon; however, this is caused not by their shorter distance, but by the moist atmosphere surrounding the earth, which intervenes between them and our sight. It is just like the apparent enlargement of objects in water, which increases with the depth of immersion.[7]

Cleomedes (*c.* early third century AD) was more specific about the refractive explanation. He was the author of a popular textbook on astronomy, said to be based largely on the views of Posidonius. Regarding the celestial illusion, he wrote:

> The sun appears to us larger when rising or setting, and smaller when overhead, because at the horizon we view it through air that is thicker and moister (for the air next to the earth has these characteristics), but through purer air when it passes overhead. Therefore, in the latter case, the ray emitted by the eyes in the direction of the sun is not refracted; while the ray directed towards the horizon, at sunrise or sunset, must be refracted because it encounters thicker and moister air. In this way the sun appears larger to us, just as objects submerged in water appear different from reality because we do not view them in direct vision [but by a refracted ray].[8]

This passage implies that refraction causes an angular enlargement of the sun. However, refraction was not clearly understood at that time – and in the next passage Cleomedes goes on to explain the sun illusion as caused by what we would now call aerial perspective, combined with size–distance invariance. We give this alternative account in Chapter 6.

These passages from Ptolemy and Cleomedes state that the horizon enlargement is similar to the angular magnification of an object seen under water. Taken in isolation, they suggest that the authors supposed that perceived size corresponds to angular size. However, in other passages both authors make it clear that perceived size corresponds to linear size through the principle of size–distance invariance. Indeed, in his later book, known as the *Optics*, Ptolemy explains underwater linear enlargement as due to angular magnification combined with the same distance as in air: 'For this reason, then, objects that are submerged in water must invariably appear larger than they would if they were observed according to direct vision at the same distance and in the same disposition.'[9]

Ptolemy did not have a complete understanding of the location and shape of the virtual image in this situation, and his statement that the magnification increases with the depth of immersion is obscure.[10] We discuss some possible interpretations later in this chapter.

A major difficulty with the refractive explanation is that enlargement occurs only when the object is in an optically denser medium than the observer. When the observer is in the denser medium, objects will subtend a *smaller* angle than usual. When looking at the moon, the eye is in a slightly denser medium (air) than the moon (empty space), and refraction should cause the moon to appear smaller rather than larger. The refractive index of air is only about 1.0003 at sea-level, while that of empty space is (by definition) exactly 1.0, so the effect is in fact extremely small when looking straight up. Ptolemy may not have been aware of this difficulty when he wrote the *Almagest*, but in his *Optics* he clearly recognized the role played by the density of the different media.[11] His knowledge may have improved between the two books. Alternatively, in the *Almagest* he may simply have been repeating the standard accounts in the scientific literature, without intending the analogy with refraction in water to be taken very literally. Like Cleomedes, he may have intended an analogy that was more like aerial perspective. In any case, he cannot have been totally satisfied with the refraction explanation, because he advanced a different explanation for the illusion in the *Optics* (described in Chapter 11).

The Arab physicist Ibn al-Haytham (eleventh century) wrote a *Commentary on the Almagest*, in which he interpreted Ptolemy as drawing a strict analogy between the effect of atmospheric vapour on the apparent size of stars and that of 'refraction' (though neither author used the word in that context) on the appearance of objects immersed in water.[12] In this work, Ibn al-Haytham took the analogy to be valid, and produced an obscure 'proof' that objects appear larger the more deeply they are immersed in water. In a later work, *On the appearance of the stars*, Ibn al-Haytham pointed out that this explanation was contradicted by the account of refraction in Ptolemy's *Optics*: he proposed instead that the atmospheric vapours formed a curved lens between the observer's eye and the celestial object, and thus produced magnification.[13] Curiously, he did not dismiss Ptolemy's account, but argued that in the *Almagest* Ptolemy was merely making an analogy, while in the *Optics* he was giving a precise explanation of refraction. He wrote:

> Furthermore, Ptolemy did not liken the magnification of the stars on the horizon to [objects] dropped in water because of what he showed in his book on *Optics*, but rather he likened [the stars] to what is dropped in water because the matter is found to be so. I mean that every visible object dropped in water is seen to be larger than when it was in air, and the farther it sinks in the water the larger its form becomes … Since Ptolemy has argued from something attested by a matter of fact, and since the stars are only seen on the horizon behind the moist vapour, then no doubt should arise with regard to this statement.[14]

Ibn al-Haytham attempted to use the correct direction of refraction to explain atmospheric enlargement, by stating that the eye is in the thinner medium and the object in the denser: 'Now if the visible object is in fog or dust, and the eye is located in thin air, sight will perceive the object's size to be larger than it really is, in the same way as it perceives objects in water.'[15] He believed that atmospheric enlargement could be a contributing factor to the celestial illusion, in addition to changes in perceived distance (Chapters 8 and 9).

A logical objection to any form of refractive magnification was raised by Wallis in 1687. He claimed that the theory implies that the angular enlargement will be present in all horizontal directions. Indeed, Ptolemy supports this view in a different context (*Almagest*, Book 9, Chapter 2): he discusses the difficulties of determining the relative positions of the stars, and states that one of the reasons is 'because the same angular distances appear to the eye greater near the horizon and smaller near the culminations'. However, Wallis argued, the angular distance right round the horizon is a fixed 360 degrees, and it is impossible to enlarge all of this at the same time: thus the enlargement cannot be real. This argument is odd, and seems to imply that a diver wearing a facemask in water could not see any real enlargement. The flaw is to suppose that an observer sees the whole 360 degrees enlarged at the same time, when in fact he samples only a small part of it in successive views. The analogy should be more like scanning the horizon through binoculars.

The fate of the refraction theory

The refraction theory implies that the horizon enlargement is measurable – but of course no such enlargement has been found by meticulous measurements over many centuries of astronomical investigation. The theory is undoubtedly wrong: however, the authority of Ptolemy and Aristotle was such that many later writers supported it as a partial cause of the enlargement. Even when accurate knowledge of atmospheric refraction became available during the sixteenth and seventeenth centuries the theory continued to be supported. Actually it seems to have reverted to a more primitive form, in which atmospheric vapours were thought to cause the enlargement. To illustrate this trend we will take a brief look at some examples of the refraction theory over the centuries.

The Roman author Macrobius (early fifth century), writing a compilation of useful knowledge for his son, was not very clear in his explanation of the refraction theory. After remarking that delicacies in eating houses are enlarged by being displayed in glass jars filled with water, he continues: 'For certainly the sun's orb, too, appears to us to be larger than usual in the morning, because the air between us and it is still dewy from the night, so that the sun's image is enlarged, just as if it were seen reflected in water.'[16] This type of imprecise statement of the theory has survived among non-scientists up to the present.

We have already described Ibn al-Haytham's contribution, and his conclusion that atmospheric refraction might contribute to the celestial illusion. The refraction theory was accepted without criticism by the Arab astronomers al-Farghani in the ninth century and Jabir ibn Aflah in the twelfth century,[17] and by the English scholar Alexander Neckam in the early thirteenth century.[18] There is a similar explanation in a book on the heavens, written for astronomers by John of Sacrobosco (who died about 1250): 'in winter and the rainy season vapours arise between us and the sun or other star. And, since those vapours are diaphanous, they scatter our visual rays so that they do not apprehend the object in its true size, just as is the case with a penny dropped into a depth of limpid water, which appears larger than it actually is because of like diffusion of rays.'[19] The scattering of visual rays sounds rather like Aristotle's idea, though the mention of enlargement in water follows the tradition of Ptolemy, Cleomedes, and Ibn al-Haytham.

Many later authors included a brief reference to either diverging rays or to enlargement in water when discussing the celestial illusion.[20] The fate of the atmospheric theory was finally decided during the seventeenth century. It seems that, despite the available astronomical evidence, the supporters of the refraction theory continued to suspect that the enlargement was real and measurable. For example, John Greaves, on his visit to Egypt in 1638, found that the setting sun near Alexandria appeared bigger than it usually seemed to be in England.[21] He measured it carefully, but found that its diameter at various times and on different days was the same. At about the same time, namely 1639, Benedetto Castelli, an Italian scientist and a pupil of Galileo, wrote his *Treatise on vision* in which he considered the possible reasons why the Big Dipper should appear so much larger when just above the horizon than when it was overhead: 'Moved by such an oddity, I wished to make sure by measuring with an instrument how much the said constellation subtends in the one and in the other posture. I always found that it subtended the same space and thus I felt sure that such a phenomenon ... of necessity had to be a fallacy of judgment and of apprehension.'[22]

Similarly, in a letter to the Royal Society in 1664, dealing with Iceland, Mr. R. Flint remarked that 'the bodies of the sun and moon seem much larger there than here; the just diameters I have not yet obtained'.[23] Even among the learned members of the Royal Society the matter was not yet settled in 1686, when on 6 April:

> a large discourse of Mr. William Molyneux was read, concerning the apparent magnitude of the sun and moon, which seem much bigger than is usual, when they are near the horizon. In it the author designed to show the absurdity of the several attempts to account for this odd phenomenon, and desired the opinion of the society thereupon. It was ordered to try, whether it be really true that the angle of the sun's diameter, when rising, is not greater than the same diameter, when the sun is considerably high.[24]

As the scientific community gradually became convinced that the refraction theory was wrong, we find signs of a return to Aristotle's 'atmospheric vapours'. The matter was discussed at a meeting of the Royal Society on 7 August 1679, when 'Mr [Thomas] Henshaw was of opinion, that the refraction of the air might cause the sun to appear bigger; and that the vapours also might augment it. Dr [William] Croune was also of the same opinion, that the vapours might augment it rather than the refraction, because the air consists of parts very different from the aether.'[25] These opinions illustrate that the refraction and vapour theories were distinguished. During the seventeenth century the latter theory was expounded in some detail by the philosopher Thomas Hobbes:

> But vision is made by beams which constitute a cone, whose base, if we look upon the moon, is the moon's face, and whose vertex is in the eye; and therefore, many beams from the moon must needs fall upon little bodies that are without the visual cone, and be by them reflected to the eye. But these reflected beams tend all in lines which are transverse to the visual cone, and make at the eye an angle which is greater than the angle of the cone. Wherefore, the moon appears greater in the horizon, than when she is more elevated.[26]

This is essentially Aristotle's explanation, except that the latter's visual rays have been replaced by rays of light. Note that there is no longer any explicit reference to refraction, which at this time was fairly well understood.

During the next century Aristotle's theory was again supported, by Samuel Dunn (1762). It even survived to the nineteenth century, in *A guide to the scientific knowledge of things familiar* by the Reverend Dr Brewer, a question-and-answer survey of science for the layman (Brewer 1863). Although he explains the horizon enlargement of the sun and moon as an illusion (p. 395), the question why hills appear larger in wet weather is answered in true Aristotelian fashion: 'Because the air is laden with vapour, which causes the rays of light to diverge more; in consequence of which they produce on the eye larger images of objects' (p. 155). This explanation therefore remained almost unchanged over a period of 22 centuries.

The difficulty of convincing non-astronomers of the illusory nature of the horizon enlargement is shown by the many statements in the literature that no measurable size change exists. For example, Molyneux (1687) described his own meticulous telescopic measurements proving the enlargement to be illusory; and Wallis (1687), when commenting on Molyneux's article, mentioned astronomers' assurances on that fact. At this time, though, there was more need to bring the point home, for in 1651 the Italian astronomer Giovanni Riccioli reported that he had found the sun and moon to be measurably enlarged on the horizon. With the help of Father Grimaldi he had measured their diameters using a sextant, and claimed that he had found values of up to a degree for the sun and 40 minutes of arc for the moon near the horizon (Molyneux 1687). These claims were never taken very seriously, but they nonetheless had to be countered.

Proof from photographs

The horizon enlargement of the moon is so compelling that its illusory nature has to be pointed out repeatedly. The constant image size of the sun and moon can be demonstrated by photographs in addition to quoting astronomical measurements. Photographs were mentioned in this connection by Haenel in 1909. These illustrative devices sometimes inspire more confidence than the most accurate measurements. For example, following Luckiesh (1921), Hamilton (1964, p. 490) declared that 'since the invention of the camera, it has been proven that the moon is no larger at the

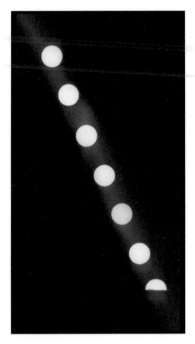

Figure 5.2 The sun rising in a very clear atmosphere over a slightly elevated grass-covered horizon at a distance of about 500 m. The photograph was obtained with a 200 mm telephoto lens, stopped down to an f-ratio of 22, through a green glass filter with a transmission of about 1/4000. The same frame was exposed seven times at intervals of about 4.25 minutes, during which the sun rises slightly more than twice its own diameter. Owing to the elevation of the horizon and the height above sea-level (about 1450 m) the lowest fully visible image is only very slightly flattened – it is just noticeable when the photograph is turned through 90 degrees. (Photograph by Cornelis Plug.)

horizon'. This type of evidence was later published by Solhkhah and Orbach (1969) and by Walker (1978a). It consists of a print from a single negative, on which the rising moon or sun has been repeatedly photographed while keeping the camera in the same position, thus recording a series of (equally sized!) images over a period of time (Fig. 5.2).

Unfortunately the camera can also lie. Many postcards are on display showing grossly enlarged photographic images of the sun or moon. These pictures are usually taken with a telephoto lens, and show a distant horizon with the moon subtending the same angular size (half a degree) as a building on the skyline. The reason the pictures are misleading is that they show no foreground, and therefore appear to represent a close shot of near buildings but with the moon subtending the same angular size as a near building. The telephoto picture of the moon and standing stones shows this effect (Fig. 1.1). Alternatively, if a standard lens is used and the foreground is shown to scale, a separate image of the enlarged moon can be superimposed on the horizon scene (Fig. 5.3). The existence of such photographs may be one reason why a gullible public still believes the enlargement to be real.

Figure 5.3 Townscape with artificially enlarged moon. (Photograph of Stirling houses by Rodger Lyall, telescopic photograph of moon by Cornelis Plug, computer combination by Peter Hucker.)

The perceptual effects of optical magnification

Many early authors suggested that atmospheric refraction causes magnification similar to that seen when looking from air into water. Though that argument is wrong, it is interesting to consider the perceptual effects of underwater magnification. It is commonly observed that an object in water viewed from air (as through a diver's facemask) appears larger than it would in air; however, the explanation of this phenomenon is by no means simple and involves aspects of both optics and perception. The different optical densities of air and water cause refraction of light rays at the boundary between the two media. The effect is to reduce the underwater part of the optical path between the eye and the object to about three-quarters of the actual distance (Fig. 5.4). The magnitude of the reduction is determined by the ratio of the refractive indices of air and water, which is about 1.00 to 1.33, or 3:4. The reduction in the optical path is equivalent to bringing the object nearer to the eye, and it increases the object's angular size by a corresponding factor (Fig. 5.5). When the line of sight is not perpendicular to the water surface the matter is much more complicated,[27] and when viewing large objects distortions of shape will occur.[28]

The increase in angular size is often presented as the sole explanation of the enlarging effect of water. However, if an object is brought closer to the eye in air, thereby increasing its angular size, this does not usually increase its perceived linear size – perhaps because the observer normally compensates for the reduced distance. According to the classical size–distance invariance hypothesis, no underwater linear enlargement should be perceived when the perceived distance corresponds to the optical foreshortening: any perceived linear enlargement should therefore be due to overestimation of the optical distance. Experimental evidence gives partial support to this view, though the overestimation of size is less than would be predicted from the overestimation of the optical distance.[29]

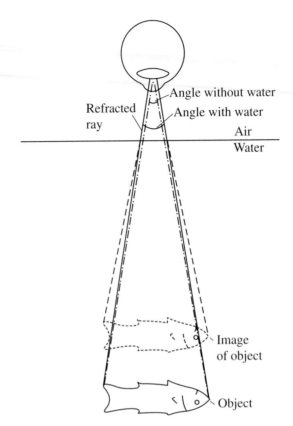

Figure 5.4 Schematic representation of the angular enlargement of an object due to its being covered by water. For small objects the enlargement is about 4/3, because that part of the optical path which is in water is reduced to about 3/4 of its length in air.

Early writers repeatedly stated that the real or perceived magnification increases with the depth of immersion of the object. There are two reasons why this might be true. The first reason is optical: if the eye is some distance from the surface of the water and the object is at a shallow location in the water, the unrefracted rays in air significantly reduce the total magnification effect; while if the object is deeply immersed, the relative contribution of the rays in air is slight and the magnification is maximal. The second reason is perceptual: water is less clear than air, and the scattering and attenuation effects of the particles produce enhanced aerial perspective (Chapter 6). The result is that perceived size and distance increase with distance over and above any optical effects.[30] Since both refraction and aerial perspective contribute to perceived enlargement in water, it is not surprising that early authors confused these effects.

Optical magnification also occurs for objects seen through field glasses and telescopes. In this case, as in water, the perceived size is less than the optical magnification might suggest. Early authors claimed that size–distance invariance was maintained, objects appearing nearer rather than larger. More recently Ronchi (1957) made a

Figure 5.5 Enlargement of an object under water. The object – an imitation fish painted on a few pebbles – was first photographed looking vertically downwards, on the bottom of an empty tank at a distance of 465 mm (top). The tank was then filled with water to a depth of 375 mm and a second photograph was taken from precisely the same position (bottom). Owing to the shortening of the optical path by the water the camera had to be refocused and the image of the object is enlarged. The expected enlargement was calculated to be about 25 per cent, but the photograph shows an enlargement of 30 per cent. (Photograph by Cornelis Plug.)

similar claim. This effect has been reported since the early days of telescopes (e.g. Wallis 1687, p. 326). However, it was not until many years later that it was considered relevant to the moon illusion. Loiselle (1898) claimed that through a telescope the moon appears closer, but of the same size (or even smaller) than with the unaided eye; but when it is simultaneously observed with the unaided other eye, the telescopic moon is seen to be larger. The failure of telescopes to magnify the perceived size of the moon was also reported by William James (1890, Vol. 2, pp. 92–3) and later by Fröbes (1923). However, anecdotal evidence is not a reliable guide to what is seen. The first author to conduct an experiment was Thouless (1968). He measured 'phenomenal' size and distance through binoculars and found that objects appeared to be slightly *reduced* in size, and to be located at about twice their optical distance.[31] Koffka (1936, p. 278) commented that the effect of looking through binoculars was like minified vision, the perceived size being so much less than the image size would predict. Size and distance relationships thus remain as obscure through binoculars as through facemasks in water. Neither type of magnification can help to explain the moon illusion.

Refraction and the oval sun

During the seventeenth century knowledge about atmospheric refraction improved substantially, so that atmospheric theories of the illusion had to be abandoned or modified. One such modification kept refraction as an indirect cause of the illusion. In order to understand and evaluate it we must briefly consider some of the ways in which the earth's atmosphere does affect the appearance of the sun and moon.

The main effect of refraction is to change the position rather than the size of the celestial bodies. Light from any source outside the atmosphere changes direction slightly when passing through the atmosphere at a shallow angle, as shown in Fig. 5.6. This causes a star, or the sun and moon, to be seen at a somewhat higher angle of elevation than if the earth had no atmosphere. The effect is quite small, amounting to about 34 minutes of arc for a star observed on the horizon at sea-level, and diminishes rapidly at higher angles of elevation.[32] As all the celestial bodies (and to a lesser extent terrestrial objects at a substantial distance from the observer) are equally affected, this physical illusion is not very noticeable. In fact, it became generally known to astronomers only from the end of the fifteenth century, when it was observed and explained by Bernhart Walther[33] – although Ibn al-Haytham and some others had been aware of the effect centuries earlier.[34] Atmospheric refraction increases the angle of elevation of a celestial body on the horizon by a little more than the angular diameter of the sun or moon, so these bodies are fully visible at a time when, without an atmosphere, they would be completely hidden from us. This 'looking over the

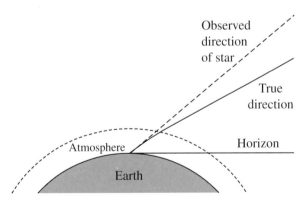

Figure 5.6 Refraction in the earth's atmosphere increases the observed angle of elevation of a star above the horizon. The effect is much exaggerated in the figure.

horizon' was discussed by Cleomedes around the third century AD, though presumably just as an educated guess.[35]

The slight raising of the celestial bodies near the horizon has an interesting side effect. The lower part of the moon, sun, or constellation is raised slightly more than the upper part, since it is viewed through more atmosphere; a distortion of shape therefore occurs, which is obvious in the otherwise perfectly round sun or moon. Only the vertical dimension of the disc is affected: it is slightly shortened, causing the sun and moon to appear oval (Figs. 5.7 and 5.8). However, its shape is often irregular, being distorted further by the effects of temperature gradients in the atmosphere[36] (see Fig. 5.9).

The flattened sun was first described by Christoph Scheiner, a Jesuit who clashed with Galileo over precedence for the discovery of sunspots.[37] Scheiner observed the flattening at Ingolstadt in Bavaria in September 1612.[38] He projected an image of the sun onto a piece of paper, and found by measurement that the vertical diameter was about 23 per cent less than the horizontal diameter. Although James Gregory regarded it as a well-known phenomenon in 1668,[39] others made special efforts to observe it for themselves. In 1675 Lucus Debes in the Faroe Islands and Bishop Gislavus Thorlocus of Iceland tried to observe the oval sun at the request of the Royal Society of London.[40] Thomas Henshaw, whose opinion on the moon illusion was quoted earlier, observed it in 1674 while on his way to Denmark. He judged the vertical diameter to be about three-quarters of the horizontal diameter,[41] remarking that he had often looked for it in England, but never could be sure of the flattening.[42]

Scheiner used this somewhat elusive phenomenon to explain the sun illusion. He thought that the contraction of the sun's vertical diameter might be perceived as an enlargement of the horizontal diameter, if the observer inadvertently treats the diminished vertical as a standard of comparison.[43] This explanation is of course not primarily based on physical principles. The real flattening of the sun due to refraction is supposed to give rise to an error of size judgement, which is the basis of the illusion. Scheiner, as a careful observer of the sun's disc, knew that the perceived enlargement was not real.

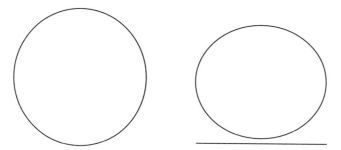

Figure 5.7 The flattened shape of the disc of the sun at sea-level due to atmospheric refraction, calculated from tables in the *Nautical Almanac*, assuming an air temperature of 10°C, a pressure of 1010 hectopascals, an undistorted angular diameter of 32 minutes of arc, and the lower limb of the sun being observed one minute of arc above the true horizon. The reduction in the vertical diameter amounts to about 16 per cent under these conditions. The disc on the left is undistorted.

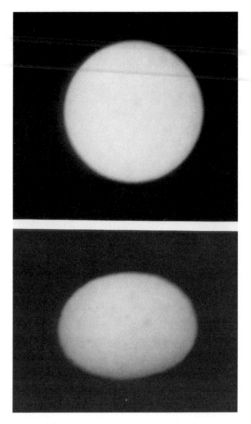

Figure 5.8 The oval setting sun. These photographs were obtained using a reflecting telescope with a focal length of 1140 mm. The aperture was reduced by screens to an effective f-ratio of 100, and in addition a neutral density filter with a transmission of 1/200 was used. The upper photograph shows the elevated and undistorted sun for comparison. The extensive flattening of the sun in the lower photograph was obtained at a specially selected site – on a west-facing escarpment on the top of Vanrhyns Pass in South Africa, at an elevation of 750 m and with a sea horizon at more than 100 km. The sun was about to set behind a low bank of sea mist, through an otherwise clear atmosphere. Its vertical diameter is reduced to 76 per cent of the horizontal diameter, and the lower edge is also visibly flatter than the upper edge. The horizontal diameter is exactly the same size as that of the undistorted upper sun. (Photograph by Cornelis Plug.)

Whereas earlier atmospheric theories treated the enlargement as real, Scheiner's theory treated it as perceptual. This amounts to a redefinition of the problem, shifting its study from the field of astronomy to psychology. However, as we mentioned in Chapter 1, this shift had already occurred centuries earlier.

Scheiner's modified refraction theory has not been actively supported, and has in fact rarely been mentioned. The reasons for this lack of enthusiasm are not hard to find. First, the difference between the horizontal and vertical diameters of the sun is too small to explain a perceived enlargement of perhaps two or more times. Second, few people have observed the oval shape of the sun, because it is only noticeable quite close to the horizon, where it is often masked by other distortions. Third, there is no independent evidence that the diminished vertical diameter can cause the sun to be seen as larger. It seems more plausible that the flattening, which represents a diminution of the angular area of the disc (amounting to about 16 per cent in Fig. 5.7), would cause it to appear *smaller* on the horizon. Scheiner's theory is therefore unlikely to be of much help in explaining the illusion.

Although atmospheric refraction has not so far proved very promising as an explanation, some further aspects deserve a mention. One of these concerns irregular distortions of the sun's disc. Such distortions can be caused by stratified layers of colder and warmer air through which the sun's rays pass when rising or setting. Because the refractive index of cold air is higher than that of warm air, different parts of the disc can be differently affected, resulting in irregular changes of shape. Samuel Dunn (1762, pp. 465–6) provided an early description of these:

> At sun-setting, such protuberances and indentures have appeared to slide or move along the vertical limbs, from the lower limb to the higher, and there vanishing, so as often to form a segment of the sun's upper limb apparently separated from the disc, for a small space of time. At sun-rising, I have often seen the like protuberances, indentures and slices, above described; but with this difference of motion, that at sun-rising they first appear to rise in the sun's upper limb, and slide or move downward to the lower limb; or, which is the same thing, they always appear at the rising or setting of the sun, to keep in the same parallels of altitude … These protuberances and indentures … enabled me to conclude, that certain strata of the atmosphere, having different refractive powers, and lying horizontally … must have been the cause.

Much later, Henning (1919) returned to these appearances and described them in great detail. Although not wishing to revive the refraction theory, he thought that these distortions might contribute to the sun illusion in a small way. He obtained sketches of the distorted rising and setting sun from several observers, who reported that the distortions sometimes cause the sun to appear larger. Some of these are shown in Fig. 5.9. In sketch (b) the split disc could be seen as a somewhat larger disc which is partly obscured, whereas in (c) the vertical dimension of the sun is momentarily

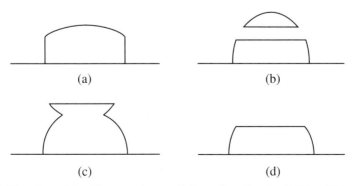

Figure 5.9 Distortions of the rising or setting sun. Redrawn from Henning (1919, p. 293).

measurably enlarged by the distortion. Similar effects can of course be seen for terrestrial objects in the form of mirages.[44]

These drawings were all made when the sun projected only partly above the horizon, introducing a configuration that may in itself cause perceived enlargement. In sketch (d) for example, the flattened upper edge of the sun could create the impression of a larger disc of which only a small part is visible, as illustrated in Fig. 5.10. This idea was taken up in more recent times by Solhkhah and Orbach (1969), who regarded it as a factor that could sometimes contribute to the illusion. They argued that two illusory mechanisms were involved. First, the upper edge of the disc is flattened by refraction. Second, an illusion of curvature can play a role – smaller segments of a disc seem to have less curvature than larger ones and are consequently judged to belong to a larger disc (Fig. 5.11).[45] Both mechanisms cause the upper edge to appear increasingly flatter as the sun sinks below the horizon, thus possibly causing an increasing perceived enlargement as the sun disappears.

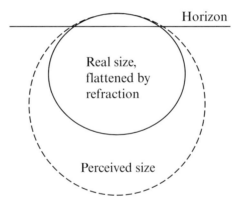

Figure 5.10 The setting sun may be perceived as too large because the visible upper edge is flattened by refraction, and because a small segment of a curve tends to appear too flat due to a geometric illusion of curvature.

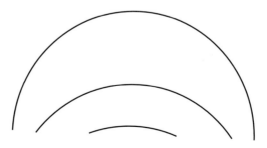

Figure 5.11 An illusion of curvature. All three arcs are parts of the same circle, but the shorter ones appear to have a much larger radius of curvature.

This partial explanation of the illusion is both elegant and plausible, but its applicability is limited. It cannot apply when the sun's entire disc is visible, as is usually the case; and it cannot explain the perceived enlargement of the constellations.

We have now excluded refraction as an important contributory factor in the celestial illusion. In the next chapter we turn to other effects of the atmosphere such as changes in brightness, colour, and contrast.

Summary

Many people today still regard the horizon enlargement as real and caused by atmospheric refraction. This erroneous explanation developed from Aristotle's rudimentary ideas on atmospheric vapours (fourth century BC). The refraction theory was refined by Ptolemy (second century), Cleomedes (*c.* third century), and Ibn al-Haytham (eleventh century) who likened the celestial illusion to the enlargement of objects seen through water. Many early authors confused refraction with vapours, or with aerial perspective. Accurate knowledge of refraction became available in the sixteenth and seventeenth centuries, but the refraction theory was only gradually abandoned and vapour theories were resurrected. Optical magnification through water and telescopes does not produce the degree of perceptual enlargement predicted by optical theory. Refraction can explain shape distortions of the sun and moon on the horizon, but not enlargement.

....

CHAPTER 6

Aerial perspectives

In the previous chapter we described how the atmospheric refraction explanation of the horizon enlargement was eventually abandoned by scientists – though it is still believed by many lay persons. One reason for this persistent belief may be the common experience that the hills and other distant objects appear enlarged in misty weather. Moreover, in certain atmospheric conditions the moon looks exceptionally large, even when quite high in the sky, and people say to each other 'Doesn't the moon look large tonight!' If refraction is not the cause, some other aspect of the atmosphere must be responsible. There are several possibilities connected with brightness, colour, and contrast. They are often loosely grouped together under the name *aerial perspective* or *atmospheric perspective*, but to clarify them we first need to define some modern terminology.

In our earlier discussions of size we were careful to distinguish between physical and perceptual definitions. Similar care must be taken with terms like brightness and colour. Whereas in the past there was little consensus among psychologists on size terminology, visual scientists have good agreement on photometry and colorimetry. *Illuminance* (or illumination) is the intensity of light energy falling upon a surface. *Luminance* is the light intensity reflected or radiated per unit area of the surface, while *brightness* (or *lightness*) is the subjective perception of luminance. Colour or hue is defined physically as a *wavelength*, or location in the visible part of the electromagnetic spectrum, with short wavelengths corresponding to blue light and long to red. The subjective perception of brightness and colour does not usually correspond to the physical measures of the reflected light, but rather to the surface properties of the object. Observers show considerable *brightness constancy* and *colour constancy*, perceiving the constant nature of the reflecting surfaces of objects despite changes in the illumination. In earlier times writers did not make all these distinctions: 'brightness' and 'colour' often included both physical and subjective measures, and 'colour' was sometimes used to include brightness. These imprecise usages will be evident in some of the earlier passages that we quote. We will also use these terms in a general manner when it is unclear what was measured or intended.

The physics of aerial perspective

The earth's atmosphere is not perfectly transparent, because it contains air molecules and other minute particles and water droplets that scatter and absorb some of the light passing through it. This affects the appearance of distant objects in several ways. The most important of these is a reduction in contrast between the object and its background with viewing distance. The contrast drops for both luminance and colour, making distant objects appear faint and hard to distinguish. In addition to the loss of contrast, there is also a change in absolute luminance and colour, with objects usually becoming lighter and bluer in the distance. We shall discuss the contrast effects first, and consider absolute luminance and wavelength later in this chapter.

The understanding of visibility through the atmosphere is much more recent than that of refraction. Bouguer developed the first quantitative theory in the eighteenth century, based on measurements of the reduction of the moon's luminance by the atmosphere at various elevations.[1] More accurate theories could not be developed until precise measuring instruments became available in the twentieth century. It is now known that contrast is reduced because some of the image-forming light from distant objects is absorbed and scattered, while other light is also scattered towards the observer, adding 'air light' to the whole scene.[2] The latter effect is the more important, causing distant objects to take on the same colour and luminance as the sky background. In a clear atmosphere this colour is blue, but in an industrial smog it is yellow, and in a clean mist it is white. For an object to be visible it must have some luminance or colour contrast with its background, luminance contrast being more effective. The luminance contrast of a black target against a sky background drops off exponentially with distance (rapidly at first and then more slowly), and the distance at which the contrast is reduced to 2 per cent is known as the visual range. The form of the function is shown in Fig. 6.1 for various ranges of atmospheric visibility.

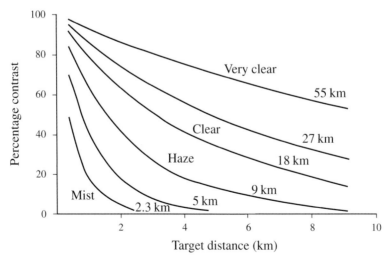

Figure 6.1 Reduction of luminance contrast with viewing distance under different atmospheric conditions for a black object seen against a sky background. The visual range is the distance at which contrast drops to 2 per cent. (Redrawn from Fry, Bridgman, and Ellerbrock 1949).

Experimental work on the effects of aerial perspective on perceived distance is fairly recent. Given the systematic effects of viewing distance on luminance contrast, it can be predicted that objects of low contrast with their background should appear further away than objects of high contrast. This has been confirmed by experiments in which contrast was measured.[3]

Aerial perspective in the early literature

Aerial perspective is an obvious phenomenon that was noticed in the distant past, and landscape painters often simulate it to create a realistic impression of distance. Ptolemy (second century) was well aware of this: 'Thus mural-painters use weak and tenuous colours to render things that they want to represent as distant.'[4] Leonardo da Vinci (1452–1519) is said to have introduced the term 'aerial perspective'.[5] He certainly discussed the topic in his *Notebooks*: 'There is a kind of perspective which I call aerial, because by difference in the atmosphere one is able to distinguish the various distances ... Because of distance we shall lack perception of parts and outlines.'[6]

Aerial perspective is often mentioned for its effects on perceived size, in addition to perceived distance. The physiologist Galen, a contemporary of Ptolemy and fellow citizen of Alexandria, held that colour indicated the spatial relationships of objects at various distances, and that size could be judged from colour: 'Vision extends right through the air to the colored object. Recognition of size and form are based only on the color of the object. No other sense perception can perform this function with the occasional exception of tactile perception.'[7] The subject was later mentioned by Plotinus (*c*. 205–70), who lived in Alexandria and then in Rome. He noted the correlation between faint colouring and the small perceived size of distant objects, whilst dismissing the former as a cause of the latter:

> Or again, it may be that magnitude is known incidentally [as a deduction] from the observation of colour. With an object at hand we know how much space is covered by the colour; at a distance, only that something is coloured, for the parts, quantitatively distinct among themselves, do not give us the precise knowledge of that quantity, the colours themselves reaching us only in a blurred impression ...

> Still the colours seen from a distance are faint, but they are not small as the masses are.

> True, but there is the common fact of diminution. There is colour with its diminution, faintness; there is magnitude with its diminution, smallness; and magnitude follows colour diminishing stage by stage with it.[8]

Plotinus was concerned with the *diminution* of perceived size with distance, which he thought must be due to something other than a reduction in angular size: his preferred explanation was the lack of a detailed relative size scale (described in Chapter 3). However, most other authors have used aerial perspective to explain the *enlargement* of perceived size compared to angular size.

Aerial perspective and size–distance invariance

As we will show, there are numerous reports that objects look larger in misty conditions; but the reasons given are as controversial as those for the celestial illusion. Classical size–distance invariance predicts that perceived linear size should increase systematically with perceived distance (which increases as the luminance contrast decreases). Cleomedes (*c.* third century AD) gave this type of explanation for the sun illusion, though he did not clearly differentiate it from his refraction explanation (described in Chapter 5). In the following passage Cleomedes gives the change in perceived distance as the primary explanation, rather than any change in angular size:

> However, the sun's distance seems to us to vary. The sun appears to us nearest in the meridian, and further away when rising and setting, and it appears even more distant when viewed from the top of the highest mountains. Whenever it appears near it appears very small; and whenever its distance appears greater, its size also appears larger. The quality of the air is the cause of all such things. In effect, when we view the sun through humid and dense air, it appears to us larger and more distant, while through pure air it appears smaller in size and at a nearer distance. Therefore, said Posidonius, if it were possible for us to see through solid walls and through other bodies (as the myth of Lynceus relates), the sun would appear to us much bigger and situated at a much further distance when viewed through these objects.[9]

Cleomedes ascribes the perceived distance explanation to Posidonius, and mentions the latter's jocular example concerning Lynceus. In Greek mythology, Lynceus was one of Jason's Argonauts who acted as look-out: he was sharp-eyed as a lynx, and could reputedly see through rocks or trees or earth. Then, as now, tall stories were used as teaching devices.

Ptolemy gave a similar explanation for aerial perspective (though, unlike Cleomedes, he did not use it to explain the celestial illusion):

> The same illusion also stems from differences in colors, for an object whose color is dimmer seems farther away and is therefore immediately assumed to be larger, just as happens with objects that actually are – i.e., when objects are seen under equal angles while some of them lie at a greater distance.[10]

Leonardo da Vinci seems to have believed that aerial perspective could explain the sun illusion. He denied that the latter was due to refraction, and added: 'for the things seen through the mist are similar in colour to those which are at a distance, but as they do not undergo the same process of diminution, they appear greater in size'.[11] It is unclear from this passage whether Leonardo thought that the size enlargement in aerial perspective was an example of size–distance invariance, or of relative size or expected size.

Most later authors backed classical size–distance invariance as the explanation. For example, Le Cat (1744) wrote: 'mists on the horizon also make us see the moon obscurely, as though it were at twice its distance. These same mists do not diminish the size of the moon's image, and since my mind has no notion of the actual size of this planet, it judges it twice as large.'[12]

Similarly Thomas Reid wrote in 1764 that in a fog he might mistake a seagull at 70–80 yards for a man on horseback at half a mile.[13] Burnet (1773, pp. 30–1) spelt it out in detail:

> Another proof of this [size–distance invariance] is what happens when we are deceived with respect to the distance, as when we see things through a fog: for from the dimness of the image upon the retina, we infer, that the object is at a considerable distance; and from this supposed distance, compared with the greatness of the image upon the retina, we conclude, that the object is much greater than it truly is. And in this way, a dog seen through a mist appears as big as a horse, and an ordinary man looks like a giant.

This common view is well illustrated by an account in a London daily newspaper during the mid-nineteenth century concerning a thick fog, which 'has seldom been exceeded in opacity in the metropolis or its neighbourhood. To see with any distinctness further than across the street was impossible.'[14] The account then describes the effect of the fog on size perception:

> On the Thames, as on land, the tendency which fog has to enlarge distant objects was strikingly illustrated; the smallest vessels on their approach seemed magnified to thrice their usual dimensions. St Paul's had a prodigious effect through the mist, though neither that nor the Monument was visible above the height of the houses. This optical illusion is said to arise from the fog diminishing the brightness of objects, and consequently suggesting a greater distance; since while the visual angle remains the same, the greater the distance, the greater the real magnitude.

Helmholtz,[15] and many other authors, stated that the moon seen through a haze appeared further away and larger. They took the view that aerial perspective was at least a contributing factor to the moon illusion, and had its effect through increased perceived distance.

Dissenters from size–distance invariance

There were a few dissenting authors. Ibn al-Haytham (writing about 1040) was unclear on the matter, and confounded several possible elements of aerial perspective. He stated that perceived distances in a fog were uncertain owing to the absence of ground cues, and could be exaggerated or diminished. Fog could enlarge objects through refraction (if the eye was in a thinner medium), and enlargement could make objects appear closer:

> Sight cannot make sure of the magnitude of the distance of an object seen through fog or dust … especially if the object lies in fog or dust while the eye is located in thin air. Sight … may believe the object to be far when it is in fact near … Or sight may believe the visible object to be near when it is in fact far, as in the case of mountains and hills … That is because of their large form; for when sight perceives [their] large size it fails to perceive correctly the intervening surface of the earth that is close to the eye, since their large size is due only to their closeness … Now if the visible object is in fog or dust, and the eye is located in thin air, sight will perceive the object's size to be larger than it really is, in the same way as it perceives objects in water.[16]

Others have equivocated about perceived distance in a fog. For example, Abbott (1864, p. 43) wrote: 'It will be found, too, I believe, that brightness is often conjoined with an increase in the apparent distance, and dimness with a decrease ... When a mist is so dense that the apparent magnitude of a mountain is increased, instead of appearing farther off it seems nearer, sometimes almost overhanging.' This description is reminiscent of a supposed real angular enlargement, perhaps similar to Aristotle's 'atmospheric vapours', as was argued by Abbott's contemporary, the Reverend Dr Brewer (Chapter 5).

Another dissenter was Myers (1911, pp. 282–3), who wrote: 'But this [size–distance invariance] explanation is quite inadequate; for, owing to the exaggerated size of familiar objects seen through a fog, they appear to us as nearer, not further, than they actually are.' However, Myers never suggested that the image size was enlarged. He went on to attempt to rescue size–distance invariance by invoking the idea of an unconsciously registered far distance which determined the perceived size. We describe this idea in more detail in Chapter 9 as the 'further-larger-nearer hypothesis'.

A different explanation – similar to some of Ibn al-Haytham's suggestions – may lie at the root of the reports of close appearances in misty conditions. Objects are sometimes described as 'looming large' in a fog: in very dense conditions objects do not become visible until they are quite close, and their angular size then increases rapidly as the observer approaches them. The fog reduces the visibility of other ground cues that normally assist in the scaling of size and distance, so the increasing angular size becomes a compelling cue to close distance.[17] This idea is similar to the 'loom-zoom' account of the moon illusion, developed by Hershenson (1982, 1989a), which we describe in Chapter 10. The loom phenomenon may help to explain the perceived closeness of objects in a mist when the distance between the object and the observer is decreasing; but it is not relevant to static viewing conditions, or to very distant objects such as the moon.[18]

There appears to be little experimental evidence on the validity of classical size–distance invariance in aerial perspective.[19] In an outdoor experiment, Ross (1967, 1975) used six observers in a thick fog and 19 in clear weather. The observers made numerical estimates of the distances and linear sizes of various white discs (Fig. 6.2): the diameters of the discs were overestimated in a fog by an average factor of 1.2 in comparison with clear weather, while the distances were overestimated by a factor of

Figure 6.2 The display of white discs of various sizes and distances in clear conditions (left) and a fog (right), as used by Ross (1967). (Photograph by Helen Ross.)

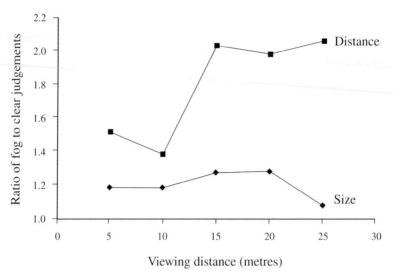

Figure 6.3 The overestimation of size and distance in a fog compared to clear conditions. Size was overestimated much less than distance. (Drawn from data in Ross 1967).

1.8, and there was a poor correlation between size and distance overestimation at different distances (Fig. 6.3). The results support a weak form of size–distance invariance, and lend no support to the idea that objects appear closer in a mist in static viewing conditions.

The experimental and observational evidence agree that terrestrial objects appear enlarged in a thick mist. Most, but not all, reports also agree that objects appear further away. It remains unclear whether the enlargement of perceived size is caused by an enlargement of perceived distance, or whether the two are independent. We now turn to Berkeley's views on the matter.

Berkeley's views on aerial perspective

A fundamental dissenter to size–distance invariance was Bishop George Berkeley, who proposed that 'faintness' was an important learned cue to both size and distance independently (1709, Sections 55–8). Berkeley noted that angular size alone was ambiguous, as a small object nearby and a large object at a distance could have the same angular size; faintness, however, could be a useful cue to both size and distance. Experience has taught us that distant objects appear fainter than near ones, and that distant objects are larger than their visual angle would suggest. If two objects have the same angular size the fainter one will be seen as larger, and as more distant.

The idea of faintness as a cue to size was also applied by Berkeley to the moon illusion:

Now, between the eye and the moon, when situated in the horizon, there lies a far greater quantity of atmosphere, than there does when the moon is in the meridian. Whence it comes to pass, that the appearance of the horizontal moon is fainter, and therefore it should be thought bigger in that situation, than in the meridian, or in any other elevation above the horizon (Section 68).

Berkeley felt that his theory also explained the variability of the illusion:

> Further, the air being variously impregnated, sometimes more and sometimes less with vapours and exhalations fitted to return and intercept the rays of light, it follows, that the appearance of the horizontal moon hath not always an equal faintness, and by consequence, that luminary, though in the very same situation, is at one time judged greater than at another (Section 69).

> Add to this, that in misty weather it is a common observation, that the appearance of the horizontal moon is far larger than usual, which greatly conspires with, and strengthens our opinion (Section 71).

Berkeley's explanation of the moon illusion formed a substantial part of his theory of vision – a theory that helped to shift opinion away from geometrical optics towards an empirical approach to perception. Yet the aerial perspective account of the illusion has received only indifferent support from others. One of the reasons for this was probably that a rival theory involving intervening objects and the perceived distance of the moon was firmly established at the time (Chapter 9). Aerial perspective therefore came to be seen as only one of several possible causes of the illusion, contributing mainly to its variability. It was mentioned as such by several subsequent authors.[20] One of the few strong supporters of the theory was the mathematician Euler, who explained it in his letters of 1762 in great detail.[21]

Criticisms of Berkeley's account

An early opponent of Berkeley's aerial perspective explanation was Robert Smith, who supported a perceived distance explanation. He raised several objections in his *Compleat system of opticks* (1738). He pointed out that the moon should appear larger when it has less brightness, but fails to do so. For example, it has less brightness contrast by day than by night; it is less bright than the sun, which has the same angular size; and when darkened by an eclipse it does not appear to increase in size. Similar points were raised by other authors.[22] The critics often failed to appreciate Berkeley's argument that faintness is only a cue to enlargement if the two have become associated by previous experience – just any faintness will not do:

> It is not faintness any how applied, that suggests greater magnitude, there being no necessary, but only an experimental connexion between these two things: it follows, that the faintness, which enlarges the appearance, must be applied in such sort, and with such circumstances, as have been observed to attend the vision of great magnitudes.[23]

Berkeley's intended meaning is fairly clear, but his use of the term 'faintness' invited misinterpretation. Among other problems, it led to a confusion between luminance contrast and absolute luminance, which can change independently of each other and which can have opposite effects on perceived size. Smith's observations concerning the brighter sun and the darker eclipsed moon can perhaps be explained by the effects of absolute luminance.

Smith also claimed, as did Filehne (1894), that aerial perspective could not explain the horizon enlargement of the constellations: these authors argued (without evidence) that the perceived size of a constellation is unaffected by the brightness of its individual stars. Filehne thought the constellation illusion could have some other cause, but he was justifiably hesitant to admit to separate explanations for illusions that seem so closely related. However (as we noted in Chapter 1), the constellation illusion is smaller than the sun and moon illusions – and a reduced role of aerial perspective could be one of the reasons.

Filehne also questioned the role of aerial perspective in the sun and moon illusions. He thought that aerial perspective would not apply to bright objects seen against a dark sky – indeed, he found that the sun often looked smaller in a mist. It is in fact a common observation that the bright sun and moon appear smaller through a thick mist, and this might seem to be damning evidence against a role for aerial perspective.[24] However, like Smith's observations on the dark eclipsed moon, the effect can be explained by a reduction in absolute luminance. As described earlier, the available evidence certainly supports a role for thick mists in the perceived size and distance of non-luminous terrestrial objects; but we know of no similar experiments for luminous targets or the celestial bodies. We have only the qualitative impressions of several authors who reported that the illusion was enhanced when the atmosphere was hazy. There is clearly a need for systematic experiments to confirm an effect of aerial perspective in the moon illusion.

We have now discussed the main aspect of aerial perspective – the reduction in *contrast* with viewing distance. We have to conclude that thick mists do not contribute to the celestial illusion, though light hazes might do so. However, haziness is also associated with changes in *perceived brightness and colour*, and either of these might contribute to the illusion. We investigate these possibilities in the remainder of this chapter.

Changes in brightness

The perceived brightness of an object can change either because its own luminance changes or because the background luminance changes. The background luminance produces a contrast effect, with objects appearing brighter against a darker background and dimmer against a lighter background. That is why the moon appears much dimmer by day than by night – because it is seen against a light sky background. As far as an object's luminance is concerned, two opposing factors are at work in daylight viewing through the atmosphere: the loss of image-forming light by absorption and scattering, and the addition of 'air light' by scattering. The result is that bright objects become dimmer, and dark objects brighter, when viewed through more atmosphere. The celestial bodies are bright, so they will always become dimmer near the horizon, whether viewed by day or night. In addition, the background sky is lighter near the horizon, and the contrast effect further reduces the perceived brightness of the celestial bodies when they are low in the sky.

The literature concerning the effects of brightness on perceived size and distance is confusing, and it is often unclear whether absolute luminance or luminance contrast was the relevant variable. We have noted that, *in misty conditions*, aerial perspective makes non-luminous objects of lower contrast appear *larger*, and usually further, than those of higher contrast. We now discuss the contrary observation that, *in some conditions*, dimmer objects appear *smaller* and further than brighter objects: this rule seems to apply to all objects viewed in clear conditions, and also to luminous objects in a mist.

Observations and explanations of the brightness effects

Observations on brightness effects are scattered sporadically throughout the literature. Early explanations were primitive and the observations were often contradictory. The enlargement of white objects was noted by Plato, who put it down to expansion of the visual ray by large incoming particles.[25] Ptolemy denied that the rays could be expanded in this manner, and also denied that brighter objects appear larger: 'The visible portion of the sky ought therefore to appear larger in daytime than at night, but nothing of this sort seems to happen.'[26] Ibn al-Haytham[27] stated that white objects appear clearer and nearer than dark objects, but did not comment on their perceived size. Leonardo da Vinci did report a size effect for luminance contrast: a dark object against a bright background appears smaller, and a bright object against a dark background larger than usual.[28] This illusion seems to have been known to the ancient Greeks.[29] It is illustrated in Fig. 6.4.

Modern explanations began to appear in the seventeenth century. Descartes noted that bright objects appear both nearer and larger than dim objects.[30] He argued that the distance effect was due to an association between pupil size, image clarity, and distance; the size effect, however, was due in part to an increase in effective image size:

> Thus the reason for their appearing closer is that the movement with which the pupil contracts in order to avoid the force of their light is so joined with the power that disposes

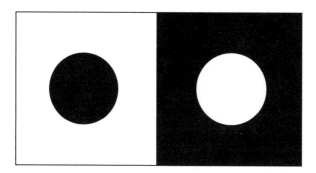

Figure 6.4 The white circle on a black background appears larger than the black circle on a white background.

the entire eye to see close objects distinctly (by which we judge their distance) that the one can hardly occur but that the other does not also occur to some extent ... And the reason why these white or luminous bodies appear larger consists not only in the fact that the estimate we make of their size depends on that of their distance, but also on the fact that they impress bigger images on the back of the eye.[31]

Descartes[32] explained the 'bigger images' by an increased retinal response to bright light: he argued that the nerve endings at the back of the eye responded to the more brilliant impulses rather than the less brilliant ones. A variant nineteenth-century account – known as *irradiation* – is that brighter images are physically larger. This idea is often associated with the name of Helmholtz, though he was predated by Plateau.[33] Helmholtz believed that irradiation was caused by blur circles in the retinal image, the effect increasing with poor accommodation. Thus a bright object on a dark background tends to spread slightly and appears larger than on a bright background. Similarly a dark object on a bright surface is encroached upon by the bright surround, and consequently looks smaller. However, modern research casts doubt on whether a small effect of irradiation could explain the variations of perceived size and distance associated with changes in luminance.[34]

A different explanation is that there is some learned association between increased luminance, enlarged image size, and reduced distance. It is not clear why this should be. Certainly there is a reduction in luminance contrast for distant objects owing to atmospheric attenuation, but dark objects actually become lighter in the distance in daylight. However, even within short distances in a clear atmosphere, bright objects appear brighter when near. It may be that the brain sums the total luminance of the retinal image, with the result that an object covering a larger retinal area appears brighter even though the amount of light per unit of solid angle is unaffected by distance.[35] The argument would then have to be that a real or perceived increase in brightness evokes an impression of enlarged size and of a closer distance. This is an unsatisfactory argument, because disappointed expectations usually give rise to contrast effects (i.e. the opposite of what is expected).

Several other authors have reported one or both of the size and distance effects in laboratory experiments.[36] Ames (1946) varied the perceived brightness of a disc of light by changing the luminance of its background: a darker background gives higher contrast and makes the disc appear brighter. The disc appears to expand and approach when it appears brighter (increased contrast), and to shrink and recede when it appears dimmer (reduced contrast). Such findings contradict classical size–distance invariance, which requires an object of constant angular size to appear smaller if it appears nearer. An alternative account is that greater perceived brightness causes an increase in perceived angular size, which then acts as a cue to closer distance. Increased luminance has been reported to increase angular size matches by a factor ranging from as little as 4 per cent (Robinson 1954) to as much as 59 per cent (Holway and Boring 1940c). The discrepancy between these authors may result from the use of only two or three observers, or from different luminance levels, or from other procedural differences.

Brightness and the moon illusion

The moon is made of a dark rock with an irregular surface, and it has a low reflectance of about 7 per cent. It is visible mainly because of the light that it reflects from the sun, which makes it appear luminous or silvery against a dark sky background.[37] However, scattered light from the earth can sometimes make the old moon faintly visible within the crescent of the new moon. In some societies this phenomenon has been considered to be a portent of bad weather,[38] as in the medieval ballad of Sir Patrick Spens:

> I saw the new moon late yestreen
> Wi' the auld moon in her arm;
> And if we gang to sea, master,
> I fear we'll come to harm.[39]

The brightness of the moon varies for several reasons. The most important of these, both numerically and for the study of the moon illusion, is *atmospheric extinction*. The luminance of the moon close to the horizon is reduced by the earth's atmosphere by a factor of 10 to 20 compared with its luminance at an elevation of 60 degrees, and by much more in comparison with higher elevations.[40] Typical values for the change in luminance with elevation for the full moon in clear weather are shown in Fig. 6.5. The perceived change in brightness is even greater than this at night, because of luminance contrast: the bright high moon is viewed against a dark sky, and the dimmer low moon against a lighter sky. The contrast with a day sky is not as great, and the moon appears much paler than at night.

The phase of the moon is the second largest factor affecting its brightness.[41] In terms of reflectance, it decreases from about 15 per cent when full, to about 8 per cent the next day, and to only 2 per cent when a quarter full.[42] The change of luminance with phase is thus most dramatic near full moon. The changes occur partly because the

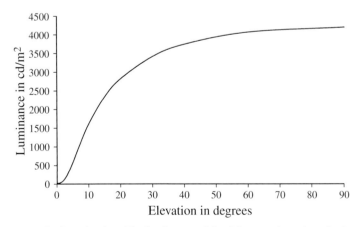

Figure 6.5 Atmospheric extinction. The luminance of the full moon in a clear sky is reduced at various angles of elevation. (Drawn from data in Conrad 2000.)

surface of the moon is irregular, and mountains do not reflect much light: the middle part of the moon reflects back more light because it directly faces us. The full moon is about ten times brighter than the quarter phase, partly because of its larger illuminated area, but mainly because it is in full opposition to the sun.[43]

Changes in brightness of 10 to 30 per cent can also occur for other reasons,[44] but such changes are not obvious when the moon is seen in isolation and has to be compared with a remembered brightness on some other occasion. However, changes by a factor of 10 or more are quite noticeable, particularly when they occur over a few hours or days. Changes are also easily detectable when they occur rapidly, as when clouds pass over the sun or moon. Shifting cloud and mists present an opportunity to observe the sun shrinking as it dims or expanding as it brightens. One of us (HER) and another observer estimated that the sun's diameter appeared to change by about 4 per cent in these conditions, and that the expanding sun appeared to approach slightly, like the front of an expanding balloon. The sixteenth-century astronomer Tycho Brahe commented on the perceived size of the moon when its luminance changes with its phase: he estimated that the faintly visible new moon appeared to be part of a disc about 20 per cent smaller than the bright full moon.[45] Similarly, the bright sickle moon appears to be part of a slightly larger disc than the faintly visible old moon.[46]

These observational reports on the effects of the moon's brightness on its perceived size have usually been confirmed by other experiments. Holway and Boring (1940c) used themselves and one other observer to match the perceived angular size of an adjustable circular disc of light (viewed in the dark and located at a distance of 31 or 61 m) to that of a standard circle of the same luminance viewed at the same distance in another direction. When the luminance of the standard circle was reduced by 99 per cent, its judged diameter was reduced by about 50 per cent. The authors added the interesting note that the full moon viewed with only one eye appeared to have its diameter reduced by about 10–20 per cent compared with binocular vision – a change which they ascribed to a reduction in perceived brightness with monocular vision.

On the negative side, Kaufman and Rock (1962a) tested eight observers with their portable apparatus (Fig. 4.2), and found that a reduction of the luminance of their horizon disc by 50 per cent had no noticeable effect on the magnitude of the illusion. The mean illusion ratio was 1.41 with the horizon disc of normal brightness, and 1.40 when it was dimmed. In a second experiment they set both discs at the zenith, but one was viewed against a normal daylight sky and the other against the sky darkened by a polaroid filter. The contrast made the latter disc appear bright and luminous. Of ten observers, six said the two discs appeared equal in size and four that the apparently brighter disc appeared larger. The lack of positive findings may have been due to insufficiently large changes in real or perceived brightness. A reduction in luminance of over 90 per cent would be needed to simulate the effect of the atmosphere on the horizon moon.

The most intensive investigation of this type was undertaken by Hamilton (1966). Noting the negative results of Kaufman and Rock, he decided to repeat their investigation more fully. An instrument similar to that of Kaufman and Rock was used, with the observers looking at a glass plate in which an artificial moon was reflected so as to appear projected onto the sky beyond. The observer viewed one disc just above the horizon and the other at elevations of 15 to 90 degrees, and then adjusted the two discs until they appeared to be of equal size. The ratio of the upper disc to the lower one gave a measure of the illusion. When the observers compared two discs at the same angle of elevation (50 degrees) with one disc only 5 per cent as bright as the other, the brighter disc was judged larger in diameter by 13 per cent. The same result was obtained against both the day and night skies. When the bright and dim discs were viewed against the night sky at different elevations, luminance was found to affect the size of the moon illusion. With the brighter disc on the horizon and the dim one at an elevation of 50 degrees the magnitude of the illusion was 1.36; with the brighter disc at an elevation of 50 degrees and the dim one on the horizon it was 1.22. In this experiment, the effect of luminance can therefore be added to or subtracted from that of position; but in the natural moon illusion, the perceived brightness of the raised moon will normally subtract from the direction of the illusion. Hamilton also noted that luminance contrast was more important than absolute luminance: the illusion increased when the luminance of the background sky was lower – that is, when the disc appeared brighter through luminance contrast. In summary, the effects of brightness oppose that of the celestial illusion, with the result that the celestial illusion is about 7 per cent less than it would be if there were no decrease in brightness on the horizon.

It has to be emphasized that the effect of luminance (and luminance contrast) on perceived size is exactly the opposite of that reported under the section on aerial perspective. In misty conditions, dim objects appear large and usually distant; but in clear conditions they appear small and far away. The observer takes the prevailing atmospheric conditions into account when using the luminance or contrast of an object as a cue to its size or distance. He can only discern the atmospheric conditions if there are other objects present and visible at different distances, giving a gradient of changing contrast. Thus aerial perspective could perhaps enhance the perceived size of a low moon seen over a rich terrain in a light haze; but a higher moon, or a moon at any elevation in a thick mist, is likely to be judged by perceived brightness alone. This interpretation lends some support to Berkeley's view that the cue of faintness must be tied to particular circumstances.

Changes in colour

The colour of the celestial bodies changes with their angle of elevation because the shorter wavelengths of light are scattered by the atmosphere more than longer wavelengths.[47] Short wavelengths are associated with a blue colour, and this accounts for the blue colour of the sky and of distant non-luminous objects, which we see as a result of scattered sunlight. Long wavelengths are associated with a red colour, which accounts for the reddened appearance of the sun and moon when close to the horizon: selective removal of the blue component of their light causes red to predominate.

Many investigators have noted the reddened appearance of the sun on the horizon and concluded that the colour enhanced its perceived size. Posidonius (second century

BC) is reported to have remarked on it (Chapter 5), and comments were frequent in the twentieth century.[48] There are also descriptions of the supposed effect of reddish light, or the presence of a low sun, on the perceived size of other objects. The reddened sun was invoked by Hans Henning (1919) as a key factor in the sun illusion, and in some related phenomena. Henning's description of the horizon enlargement of the sun was based on observations made over a period of six years at the Strasbourg Observatory by a total of 60 persons. From this point the sun could be seen, at different times of the year, rising over a wide arc of the distant Black Forest and setting behind the Vosges Mountains, with horizon distances of 30 to 65 km. Henning enquired whether the perceived sizes of terrestrial objects, such as trees on the horizon, were also affected when viewed close to the rising or setting sun. The findings are worth quoting:

> It was proved at once that the apparent sizes of trees, houses, rocks, etc. on the mountain crest increase in a very similar manner. But in addition to this astonishing appearance we find a second, which is much more surprising and above all somewhat puzzling: in the vicinity of the setting sun on the crest of the Vosges mountains, one perceives things extremely clearly, which one is otherwise altogether incapable of seeing. It is downright astonishing when, in the light of the setting sun, we suddenly see clearly houses and trees, at a distance of 50 km, which shortly before, and with more intense illumination from the sun, we could not even see with fieldglasses.[49]

Henning noted that the enlargement of terrestrial objects only occurred when the sun appeared red. Observations on the island of Texel and in the Alps confirmed the phenomenon, and also indicated that the sun need not be above the horizon – objects were also enlarged in the red light just before sunrise and just after sunset.

Similar effects were reported earlier by Samuel Dunn (1762), although he did not associate them with the colour of the light. His rambling style invites a direct quotation (pp. 466–7):

> I have often observed, with admiration, the most distant trees and bushments, which at other times have appeared small to the naked eye, but whilst the sun has been passing along a little beneath the horizon, obliquely under them, just before sun-rising, when the sun has been thus approaching towards and beneath any trees and bushments, they have grown apparently very large to the naked eye, and also through the telescope; and they have lost that apparent largeness, as the sun has been past by them.

Explanations for colour effects

A red appearance was never thought to be the main cause of the sun illusion, but its supposed contribution required an explanation. Haenel (1909) argued that red is associated with nearby objects, and blue with distant horizons; as a result, the reddened sun and moon may be perceived as closer, or larger in angular size, than usual. A different explanation involves the effects of luminance on perceived distance:[50] bright colours (e.g. red) are described as 'advancing', and dim colours (e.g. blue) as 'retiring' or 'retreating'.[51] Alternatively, the effect may be due to colour contrast or luminance contrast, red generally having higher contrast with the background than blue.[52]

It should be noted that, on these arguments, red enlargement is similar to luminance enlargement in that it contradicts classical size–distance invariance. However, unlike luminance enlargement, it could work to enhance the moon illusion.

The effects mentioned above can be seen with one eye. There is also a different colour depth effect seen in binocular vision, known as *colour stereoscopy* or *chromostereopsis*, which was first investigated by Donders in 1868. It depends on the chromatic aberration of the eye (Chapter 7). The aberration produces slightly different binocular disparities for red and blue light. Most observers perceive red in front of blue under high illumination, but blue in front of red under low illumination.[53] In quantified experiments, the difference in perceived distance between red and blue light amounts to about 6 or 7 per cent.[54] This stereoscopic distance effect is very small, and it is difficult to see how it could contribute to either the sun or moon illusion. Observers usually report that any accompanying changes in perceived size go in the correct direction for classical size–distance invariance: that is, if red objects appear nearer than blue objects they also appear smaller. Thus colour stereoscopy would have to make the celestial bodies appear further on the horizon in order to produce an enlargement of perceived size. The required distance effects could only occur if the celestial bodies appeared blue in daylight or red by night. In practice, of course, the sun is only seen in daylight, and appears red on the horizon. The moon may appear reddish on the horizon by day, and yellow at night. Binocular colour stereoscopy could perhaps contribute a little to the moon illusion at night, but can be dismissed as a general contributing factor to the celestial illusion.

Experiments on colour effects

If there is a monocular colour effect (as distinct from a brightness effect), its size is arguable. Quantz (1895) found an effect of only about 1 per cent, which he noted was too small to account for the moon illusion: his observers matched the perceived angular size of discs of different colours by adjusting their distance. Over (1962a) also used a distance adjustment technique to vary angular size, and found that the perceived size of red targets was about 20 per cent larger than that of blue targets, and their perceived distance nearer by about 32 per cent. Over believed that the effect was due to chromatic aberration, to which we return in Chapter 7.

Some other experiments on colour probably entailed the use of binocular vision. Wallis (1935) had observers discriminate between the sizes of pairs of cubes of different colours and sizes, and found that yellow appeared largest, then red, green, and blue: the largest difference in judged size was about 3 per cent, and appeared to be correlated with luminance. Bevan and Dukes (1953) had their observers make linear size matches for coloured targets out of doors, and found that the size of red cards appeared about 13 per cent larger than grey cards, whereas blue cards appeared about 7 per cent larger.

The difference between red and blue cards was therefore about 6 per cent. It is unclear whether these effects were specifically binocular, or would have occurred with monocular vision.

We now return to the experimental investigation of the role of colour in the sun and moon illusions. Kaufman and Rock (1962a) used the instrument shown in Fig. 4.2 to create artificial moons at optical infinity. In one of their experiments the horizon disc was reddened by a red (i.e. minus blue) filter on half of the trials and compared to a similar white disc viewed higher against the sky. The mean size of this illusion for seven observers was 1.37 for the white disc and 1.34 for the reddened horizon disc, and the difference was not statistically significant. Colour alone therefore did not seem to have any effect on the illusion in this experiment. However, the authors later reported from their 1959 notes that they saw a large orange moon in a haze at about 5 degrees elevation over the ocean: 'The moon was hanging like a gigantic globule in a hazy partially clouded blue sky. Because of the time of year it was still daylight when the moon appeared. All present agreed that it appeared to be unusually large.'[55] Three observers used the moon machine to match a white artificial moon in the zenith to the real orange moon on the horizon, and gave a mean illusion of 1.83. This is quite a large effect, but it may have been due to variables other than the colour of the moon: for example, it was not possible to adjust the size of the real moon, so the 'error of the standard' may have been added to that of the moon illusion (Chapter 4). However, the same subjects matched the white artificial moon to the real moon when both moons were on the horizon but the artificial moon was about 40 degrees to the south of the real moon: in this case the subjects made an accurate angular match.[56] They do not, therefore, seem to have been influenced by either the colour of the moon or the error of the standard. We are left with some uncertainty as to why the colour red should produce enlargement in some laboratory experiments but not in these moon machine experiments. Perhaps colour is more effective within the context of a hazy atmosphere or reddish light. A more extensive series of moon machine experiments is needed before dismissing any role for colour.

There is at present no experimental evidence that the colour red contributes to the moon illusion. However, there is plenty of anecdotal evidence that both occur together, and red has been shown to increase both angular and linear size matches in some terrestrial experiments. Further experiments are needed to confirm the circumstantial reports that red colouring contributes to the celestial illusion in hazy conditions.

Summary

Aerial perspective, or a reduction in luminance contrast and colour contrast for objects viewed through the atmosphere, was described by many early authors. Terrestrial objects are reported to appear larger in a mist, and usually further away. Berkeley argued that aerial perspective contributed to the celestial illusion, and that it was a cue to size and distance independently. However, a thick mist cannot contribute to the illusion, because it makes the sun and moon appear smaller owing to a reduction in absolute luminance. A reduction in perceived brightness decreases perceived size and increases perceived distance. The atmosphere reduces the luminance of the celestial bodies by a factor of 10 or more near the horizon compared to higher in the sky; and low contrast with the brighter horizon sky further reduces their perceived brightness. The effect of brightness should thus counteract the celestial illusion, and it has been shown to do so by about 7 per cent in a moon machine experiment. Red colouring is reported to increase matched angular size by up to 20 per cent – but it has not been shown to have an effect in a moon machine experiment. A thin haze might contribute to the celestial illusion when terrestrial objects provide a gradient of contrasts; and the atmospheric enhancement of red wavelengths for the low sun and moon might also contribute. Experimental evidence in suitable atmospheric conditions is needed to support these ideas.

CHAPTER 7

In the eye of the beholder

We established in Chapters 2 and 5 that the cause of the moon illusion does not lie in the domains of either astronomy or atmospheric refraction. Having ruled out physical explanations, we turned in Chapter 6 to some atmospheric effects on luminance and wavelength that may affect the observer's visual system or visual perception, and may provide partial explanations of the moon illusion. In this chapter we deal with attempts to explain the illusion in terms of some other processes taking place within the eye.

Illumination and pupil size

The French philosopher and scientist Pierre Gassendi (1592–1655) proposed that the moon illusion was due to changes in the size of the pupil of the observer's eye. He reasoned that the atmosphere caused the sun and moon to be less bright near the horizon than at higher elevations, and that the reduced illumination caused the pupil to dilate. He further argued that the perceived size of an object increased with pupil size, thus making the horizon sun and moon appear enlarged.[1]

The fact that the pupil is enlarged in darkness and contracts in bright light is relatively easy to observe. About the beginning of the tenth century the Arab al-Razi wrote a treatise on the subject entitled *On the reason why the pupil contracts in light and dilates in darkness*.[2] In Europe the phenomenon was clearly described by Leonardo da Vinci around 1500 (*Notebooks*, p. 210), and the history of some early research on the subject was described by Porterfield in the eighteenth century.[3] This aspect of Gassendi's theory is uncontroversial. However, the claim that such dilation causes a significant increase in the perceived size of an object is unacceptable. This other element of the theory did not originate with Gassendi, for it is found fully developed in the notes of Leonardo da Vinci: 'Every object we see will appear larger at midnight than at midday, and larger in the morning than at midday. This happens because the pupil of the eye is much smaller at midday than at any other time.'[4] It is not clear where Leonardo obtained this idea, which appears quite strange even in the light of medieval knowledge about vision. Certainly by Gassendi's time the idea was untenable. In 1604 Kepler gave an essentially correct description of image formation by the lens onto the retina,[5] and experience with early telescopes had shown unequivocally that image size was not affected by the diameter of the lens or lens opening. The enlarging effect of the dilated pupil therefore had to be rejected. Indeed, the Italian physician Fortunis Licetus, one of Gassendi's critics, made this point in 1640. As a result, Gassendi changed his theory, and claimed that the size changes caused by contraction of the pupil were actually too small to be observed by the unaided eye, but could be determined by accurate measurements.[6] He gave a number of examples of such size changes, but these relate to the effects of changes in luminance, as discussed in the previous chapter.

Gassendi's original theory was also rejected by the physicist Robert Hooke in 1679,[7] when answering another Fellow of the Royal Society, Dr William Croune, who supported the idea. Others, however, accepted the supposed phenomenon and tried to explain it. One such explanation was published anonymously in the 1670s in France. Its author is thought to be the Abbé Pierre Michon Bourdelot[8] who, on being pressed to defend his support for Gassendi's theory, claimed that dilation of the pupil changed the shape of the eye by making the lens flatter and the eye longer. This would cause the image of the horizon moon to be proportionally enlarged.

Bourdelot's speculations soon elicited a critical response from Molyneux (1687), who also pointed out that Gassendi's theory did not explain the enlargement of the constellations, which are hardly reduced in luminance near the horizon. Later work also failed to provide any support for the changes in the shape of the eye envisaged by Bourdelot. By the middle of the eighteenth century it therefore appeared surprising that anyone could ever have supported such a theory. Porterfield, in his extensive treatise of 1759 on the eye and vision, expressed great surprise and judged Gassendi to be 'totally ignorant' of the principles of optics,[9] an opinion which was shared and expressed in similar words by Robins (1761).

Gassendi's explanation was revived early in the nineteenth century by Ezekiel Walker.[10] He supported his opinion by observations of the images of a candle produced by lenses of various sizes. Walker's work was immediately repudiated by the editor of the journal in which it was published, who pointed out that if Walker were right, the magnification of a telescope would depend on the diameter of its object lens – a patently absurd notion. Also, a piece of paper would appear to expand and contract depending on whether it was held in the shade or the light, and a room would suddenly appear to shrink in size when turning on additional lights.[11] Interestingly enough, the last two objections had already been foreseen by Leonardo da Vinci, who countered them with the hypothesis that the perceived size of an object was judged relative to that of other objects: 'And as it [the pupil] varies its size the same object ... will appear of different sizes, although it often happens that the comparison with surrounding things does not allow this change to be discerned when you look at a particular object.'[12]

Walker's work was deservedly forgotten, except for a few brief refutations;[13] but the idea that the moon illusion is due to variations in the size of the pupil has survived. Later in the nineteenth century the astronomer Paul Stroobant (1884) found that the perceived size of an artificial moon was reduced by about 30 per cent when a bright light was shone in the observer's eye. He concluded that this was probably an important contributory factor to the moon illusion. His observation has recently been linked to a different type of pupillary theory, as will appear in the next section.

Inadequate focusing and pupil size

Porterfield pointed out that if the eye were not properly accommodated then the image of the moon on the retina would be fuzzy and slightly enlarged.[14] The enlargement is caused by a blur circle surrounding the retinal image.[15] Objects of small angular size, such as stars, are affected proportionally more than objects of larger angular size, such as the moon.[16] Moreover, since the effects of poor focusing are enhanced by a large lens opening, the enlargement would increase as the pupil expands. Thus if the horizon moon is viewed with a larger pupil than the elevated moon, and if the eye is poorly focused, then a slight enlargement would be expected. This tentative explanation of the moon illusion was also briefly considered by Mayr (1904), as an afterthought to his criticism of Gassendi. He pointed out that the enlargement due to poor focusing can at best be small, as there is no evidence that we see the horizon moon grossly out of focus.

Some indirect evidence that could be interpreted as support for the focusing theory came from an investigation by Pozdena (1909). He found that the perceived size of the moon (as represented by the size of a matching comparison disc) was always larger for one of his three subjects than for the other two. This person was slightly shortsighted, whereas the other two had normal vision. The larger illusion for the shortsighted subject could therefore have been due to the moon's image being out of focus. However, the opposite effect was later reported by Witte (1919b), who found that the perceived size of the moon is usually slightly smaller for shortsighted people. Nothing can be concluded from such studies with small numbers of subjects.

Only fairly recently has the inadequate focusing theory been examined in detail – by J.T. Enright (1975), a behavioural physiologist. He began by considering reports that when the enlarged rising moon is viewed through a narrow tube it immediately shrinks to normal size, while its perceived size higher up in the sky is unaffected. This phenomenon has been either confirmed or rejected by a large number of expert observers over a period of three centuries or more, and it remains as controversial as ever. Some observers, of whom Goüye (1700) can be taken as an example, have asserted that looking through a tube at the rising moon reduces it to its size when elevated. In commenting on Gassendi's theory, Goüye claimed that 'in spite of this dilation of the pupil caused by the darkness, if one regards the moon through a small tube of paper, one finds it smaller in the horizon' (p. 9, our translation). Yet some years earlier this same phenomenon was emphatically denied by Molyneux (1687, p. 317): 'Moreover, all variety of adjoining objects may be taken off, by looking through an empty tube, and yet the deluded imagination is not at all helped thereby.' Such contradictory observations are common: we will return to this issue when we consider the role of the visible terrain on the illusion (Chapter 9).

Enright (1975, p. 93) suspected that the illusion-destroying effect of the tube depended on whether or not it had a tiny opening: 'My own experience dictates that a viewing tube with a broad opening does not produce the described effect, although one with a small terminal aperture does ... with the open tube, the horizon moon remains large; with the small aperture, it shrinks strikingly.' To test this hypothesis Enright obtained the help of 28 observers. They were given a black paper square with a hole of 2.5 mm in it, which they could place directly in front of the eye to provide an artificial pupil. Almost all the observers, who were tested individually, reported the rising moon to appear smaller and sharper with the artificial pupil than without it. The decrease in perceived size was estimated by 19 of the observers, yielding a range of up to 55 per cent, with a mean of about 20 per cent. Unfortunately similar size estimates were not taken with the moon higher in the sky, which made the results ambiguous. Further experiments convinced Enright that the reduction in size was due neither to the masking of the horizon terrain by the screen containing the aperture, nor to the reduction in light intensity caused by the small opening, but to a combination of pupil size and poor focusing.

Gassendi's assumption that the pupil is larger when viewing the horizon moon than an elevated moon had never been directly tested. Enright therefore obtained photographs of the eyes of his volunteers when making their first and last observations of the rising moon. Measurements of pupil size on these photographs proved to be 'moderately encouraging'. Pupil diameters decreased for 22 of the 25 observers as the moon rose, although the extent of the change was in most cases only between 10 and 20 per cent. He next confirmed Stroobant's (1884) observation, that shining a light in the observer's eye caused the perceived size of the rising moon to shrink markedly. He achieved this by directing the beam of a small flashlight into the observer's eye from the side, and found that the extra light caused constriction of the pupil and also a reduction in the perceived size of the moon. Removal of the light caused the moon to return to its subjectively enlarged state. These changes in perceived size were 'scarcely noticeable, if at all' (p. 96) when the moon was high in the sky.

Although Enright considered these observations to support the involvement of the pupil in perceived size, the support seems weak. His own observations indicate that even when the moon is high the pupil is still quite large. The flashlight should have caused a considerable further contraction, but this was not accompanied by any reduction in perceived size.

Reasons for inadequate focusing

It is difficult to find a mechanism that would cause the eye to be sufficiently out of focus when looking at the moon. Three possibilities will be considered: chromatic aberration, spherical aberration, and night myopia. The first of these was pointed out by Leiri (1931). He referred to Henning's idea (Chapter 6) that the horizon enlargement of the sun and moon was due to their red colour when rising or setting, and explained the enlargement in terms of the chromatic aberration of the eye. Light of different colours is not brought to exactly the same focus (see Fig. 7.1). If the lens is focused for wavelengths in the middle range of the spectrum (those of yellow and green light), then longer wavelengths (red) will be brought to a focus slightly behind the retina and shorter wavelengths (blue) slightly in front. Chromatic aberration is not normally noticeable, because the eye automatically adjusts its focus for clear viewing in the more

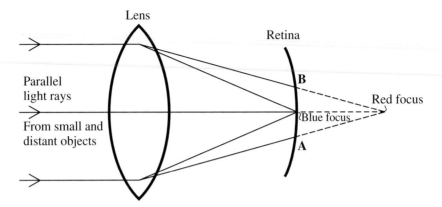

Figure 7.1 Schematic representation of chromatic aberration in an eye focused for viewing a small, distant object in blue light. The index of refraction of the lens is slightly smaller for red than for blue light; consequently the focal length of the lens is slightly larger for red light. When the lens focuses blue light on the retina the red light does not come to a focus, but is spread over a blur circle with diameter AB. The effect is much exaggerated in the figure.

intense middle range of the spectrum, and the red and blue components are usually too dim for their blurred focusing to be perceived.[17] Leiri argued that, when viewing the reddened setting sun, the eye was focused for the main components of daylight (including the blue light of the sky), causing the image of the sun (formed mainly in red light) to be out of focus. The resulting blur enlarges its image slightly, thereby causing the illusion.

While there is no doubt that the human eye is indeed subject to chromatic aberration, Leiri's explanation is far from convincing. The data on the chromatic aberration of the eye[18] imply that there is only a small enlargement of images formed by red light in an eye whose lens is accommodated to white light. The small size of the effect has also been confirmed by later research.[19] The pupil is always small when viewing the sun, thereby reducing chromatic aberration; and, in any case, the eye should accommodate to the red light of the horizon sun and thus bring it into sharp focus. Indeed, Henning's claim that he could see clearer details in red light (Chapter 6) suggests that his acuity was improved by the reduction of chromatic aberration. Furthermore, an explanation based on chromatic aberration could not apply to the enlargement of the moon and constellations at night.

A second possible explanation for Enright's results with artificial pupils is spherical aberration. We mentioned this factor in Chapter 2 to explain why bright stars appear slightly larger than dim ones. While spherical aberration might cause the horizon moon to be slightly out of focus, it could play only a very minor role in the celestial illusion. First, the extent of spherical aberration for the human eye is rather small.[20] Second, the reduction in pupil size as the moon rises (as found by Enright) is too small to cause much reduction in spherical aberration. Finally, this explanation does not account for the sun illusion (because the pupils are constricted when viewing the sun), or the enlargement of the constellations.

A third possibility considered by Enright is night myopia – a phenomenon first described by the astronomer Maskelyne (1789). In dim light the eye tends to focus at a short viewing distance, even when the observer is looking at distant objects.[21] Under low levels of illumination and after dark adaptation, the eyes of young adults have been found to focus to an average viewing distance of less than a metre,[22] although the distance varies considerably between individuals.[23] The evidence provides some support for Enright's theory. The dim horizon moon is viewed with poorly focused eyes owing to night myopia, and with enlarged pupils owing to darkness, thus causing the moon's image on the retina to be fuzzy and enlarged. When the moon is higher and brighter, both the night myopia and the pupil size are reduced, and the image-enlarging effect of poor focusing all but disappears. However, as Enright himself admits, there are serious problems with this theory too. The extent of night myopia when viewing the rising moon is unknown and may be quite small. Furthermore, there is no reason why it should be reduced as the moon rises, as the featureless extent of sky in which the elevated moon is seen can also cause myopia.[24] The theory also fails to explain the moon illusion during dusk and daylight, and cannot account for the sun illusion and the perceived enlargement of the constellations. We must conclude that the combination of variable pupil size and night myopia, as envisaged by Enright, can contribute very little to the illusion. Indeed, Enright himself (1989a, 1989b) later revoked his accommodation account, and offered instead an oculomotor account based on convergence effort (Chapter 11).

Perceived size and focusing distance

We noted that Gassendi's belief that perceived size was proportional to the size of the pupil may have given rise to the theory of pupil size and inadequate focusing. As a result of Enright's investigation this theory had to be abandoned. There is, however, a slightly different explanation that could be seen as a successor. It is no longer based on variations in pupil size, but, instead, on variations in the focusing distance of the eye. The idea is that the perceived size of an object depends on the distance to which the lens of the eye is accommodated, being smaller at a shorter focusing distance. When the moon is close to the horizon the eye focuses a considerable distance away and the moon consequently appears large. When it is in mid-sky, on the other hand, there is little for the eye to focus on (apart from the moon itself) so that it tends to lapse to its dark focus – a distance of less than a metre for most people – causing the perceived size of the moon to be reduced. It has to be asked whether the size reduction is perceptual or optical, or both. For a well-focused eye, objects of constant angular size do in theory produce a slightly smaller retinal image at very close distances. This is because the nodal point of the eye is closer to the retina with near accommodation. For a standard eye the distance of the nodal point is about 14 mm for near accommodation and about 17 mm for far accommodation,[25] giving an image reduction of up to about 20 per cent for very close objects. However, other authors argue that any optical effect is negligible.[26]

A reduction in perceived size with increased accommodation was first reported by William H. Rivers (1896). The effect was measured by Heinemann, Tulving, and Nachmias (1959), and found to be about 20 per cent – though the authors argued that the accompanying changes in convergence were the main cause.[27] These authors noted that distance judgements were not in accordance with classical size–distance invariance, since apparently smaller targets were judged further away. The size effect is usually known as *accommodative micropsia* or *accommodation-convergence micropsia* or, more generally, as *oculomotor micropsia*.[28] When focusing on a near object, the two eyes converge in order to bring the images onto corresponding points in both retinas (otherwise double images would be seen). Simultaneously, the lenses accommodate by increasing their curvature, to bring the images into sharp focus (Fig. 7.2), and the pupils constrict. When the eyes focus on a distant object their optical axes become parallel and the lenses flatten. The changes in accommodation, convergence, and pupil size normally occur together in a reflex manner, sometimes called the *near triad*.[29] Any of these changes could, in theory, serve as a cue to the distance of the object. However, because the changes are extremely small for distances beyond a few metres, they can be effective only at short distances. It should be noted that such cues are muscular rather than visual, because they are associated with the muscles controlling convergence and accommodation. Arguments exist as to whether the sensory information depends on feedback from those muscles, or on the motor command signals from the brain to the muscles, or on both (see Fig. 7.3). Regardless of its source, the oculomotor information could cause a change in perceived linear size through the size–distance invariance principle, or could

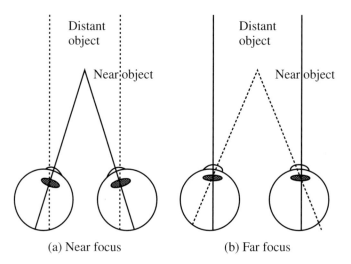

(a) Near focus　　　　　(b) Far focus

Figure 7.2 The effect of focusing distance on the retinal images of near and far objects: (a) For near focus, the eyes converge and the lenses bulge. The images of a distant object will be blurred, and it will be seen as double. (b) For far focus, the optical axes of the eyes are nearly parallel and the lenses relax their accommodation. The images of a near object will be blurred, and it will be seen as double.

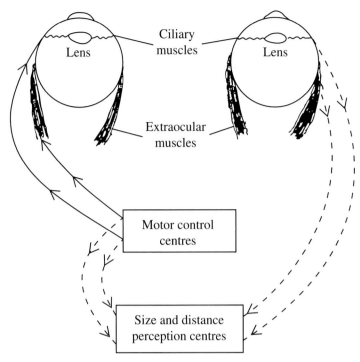

Figure 7.3 Possible explanations of accommodation-convergence micropsia. The left eye illustrates the outflow theory, in which the motor commands to the muscles are copied to the size and/or distance perception centres. The right eye illustrates the inflow theory, in which the resultant muscular tension is fed back to these centres.

affect perceived angular size independently of distance. The latter seems more likely, since micropsia is often accompanied by an increase in perceived distance. There is almost as much argument about the nature of accommodative micropsia as there is about the moon illusion. This uncertainty does not, of course, rule out accommodation as a possible contributor to the moon illusion, but it does indicate that further explanation is required.

Accommodation was suggested as the cause of the moon illusion by Roscoe in 1977, and was later tested by Iavecchia, Iavecchia, and Roscoe (1983). They used a moon machine similar to the second machine of Kaufman and Rock (Fig. 4.2) for viewing the image of an artificial moon at optical infinity against various backgrounds. Their observers first viewed the virtual moon against the sky or other less distant surface; they then viewed it at a distance of one metre against a dark background, and adjusted it to appear the same size as the remembered former moon. The size of the adjusted moon gives a measure of the perceived angular size of the first moon – but it may be a biased measure, because the locations of the standard and adjustable moons were not counterbalanced.

In one experiment six observers viewed the artificial moon at two small but unspecified angles of elevation above four horizon scenes, varying in distance from near to far. The mean perceived angular size of the moon increased by about 16 per cent from the nearest to the furthest horizon, and then declined by about 6 per cent when it was viewed slightly above this horizon. These changes are quite small, and it is not clear what the state of accommodation was. In a second experiment parts of the lower half of the visual field were screened off, leaving only a narrow strip of terrain visible in addition to the sky above the horizon. The artificial moon could be seen against the sky with either foreground (12 to 22.5 degrees below the horizon), intermediate terrain (6 to 12 degrees below the horizon), far terrain (3 to 6 degrees below the horizon), very far terrain (from the horizon to 3 degrees below it), all terrain, or no terrain visible (Fig. 7.4). The mean judged angular size of the artificial moon under each of these conditions, relative to that of the 'sky only' condition, is indicated in the second column of Table 7.1. The maximum enlargement was 26 per cent, with very far terrain visible.

The focusing distance of each observer's eye was measured with a laser optometer while they were looking at a particular scene, just before and after each of their size judgements. The mean changes in accommodation from the 'sky only' viewing condition to the other conditions are shown in the last column of Table 7.1. There is a clear relationship between the changes in accommodation and the corresponding changes in perceived angular size: as the eye focuses for larger distances (corresponding to a negative change in accommodation), perceived angular size increases.[30]

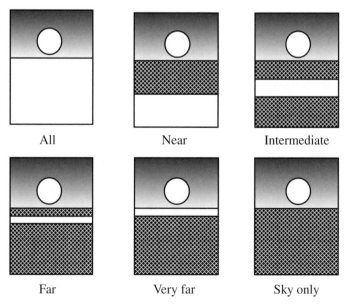

| All | Near | Intermediate |

| Far | Very far | Sky only |

Figure 7.4 Viewing the horizon moon with various sections of terrain visible. The hatched areas were screened out. (Redrawn from Iavecchia, Iavecchia, and Roscoe 1983.)

Table 7.1 Mean judged angular sizes of an artificial moon with various sections of terrain visible, and the corresponding mean changes in the accommodation of the eye for six observers. The dioptre is a unit of refracting power. 'Sky only' was used as a reference condition. (Data from Iavecchia, Iavecchia, and Roscoe 1983)

Terrain visible	Judged size	Change in accommodation
Sky only	1.00	0
Near	0.96	+0.13 dioptre
Intermediate	0.99	−0.07 dioptre
Far	1.08	−0.28 dioptre
Very far	1.26	−0.44 dioptre
All	1.25	−0.27 dioptre

The fact that perceived size and accommodation seem to be closely related does not, of course, prove that perceived size is determined by the state of accommodation of the eye. Indeed, old people with very little accommodation ability still experience the moon illusion[31] as do those without natural lenses.[32] It is possible that both are determined by some third factor. For example, in the experiment of Iavecchia and colleagues (1983) several factors varied systematically from the 'near' to the 'very far' condition (see Fig. 7.4): the distance of the visible terrain increased; the grain (fineness of texture) of the terrain changed from coarse to fine; the angle below the horizon of the terrain decreased; and the width of the strip of terrain, as well as the width of the strip of mask between the terrain and the horizon, decreased. Some combination of these factors may have been the cause of the change in perceived size. Most of these factors involve size contrast, which we shall consider in Chapter 10 as a phenomenon contributing to the moon illusion. Most are also distance cues, and could cause a change in perceived linear size through a change in perceived distance (Chapter 9). The same factors may have caused the changes in accommodation, since the grain of the visual field, the position of a stimulus in the visual field, and the size or position of a screen have all been found to affect accommodation.[33] Such considerations led Hull, Gill, and Roscoe (1982) to conclude that 'evidently eye accommodation is somehow involved in judgments of apparent size, but the nature of the involvement is far from clear' (p. 317). Although research on the problem has since continued (Roscoe 1985, 1989), the possible role of accommodation in both the artificial and the natural moon illusion remains unclear.

The optical effects discussed in this chapter are either of uncertain status or do not seem to be able to account for the celestial illusion. However, the phenomenon of oculomotor micropsia is of interest, and will be discussed further in Chapters 11 and 12 along with other motor theories of size perception. Meanwhile we shall turn our attention to psychological explanations based on the perceived form of the sky.

Summary

Changes in pupil size were rejected as a cause of changes in perceived size, and thus as a possible cause of the celestial illusion. Image blur due to inadequate focusing with an enlarged pupil was also rejected, whether caused by chromatic aberration, spherical aberration, or night myopia. Iavecchia *et al.* (1983) found that the nature of the visible terrain affected both the state of accommodation and perceived size, a very far terrain causing a perceived size enlargement of 26 per cent. The authors argued that changes in accommodation were the cause of the size enlargement; but the visible terrain could also be directly responsible, through relative size effects or through changes in perceived distance. It is unlikely that any optical changes in image size contribute to the celestial illusion, but oculomotor factors may be implicated.

CHAPTER 8

The vault of the heavens

One of the most popular theories of the moon illusion relates the horizon enlargement of the celestial bodies to the perceived form of the sky. Many people see the sky as a flattened, dome-shaped surface. If the celestial bodies are seen as moving along this surface, then the sun and moon should appear closer at higher angles of elevation than towards the horizon. But a close object of a given angular size is smaller than a distant object of the same angular size. Therefore the sun and moon should appear smaller overhead than on the horizon.

A clear illustration of this idea (Fig. 8.1) was given by Desaguiliers (1736a), and a more picturesque version (Fig. 8.2) by Robert Smith in 1738. These figures differ in that Desaguiliers shows the perceived distance of the sky as coinciding with the 'true' semicircle at the zenith and as being further than the semicircle at the horizon, while Smith shows the reverse. It is a version of Smith's figure that appears in almost all later publications. Thomas Young (1807) was unusual in giving a version of Desaguilier's figure, in which the flattened dome is perceived as further than the true semicircle at all points (Fig. 8.3). Young's figure has only occasionally been reproduced.[1]

The flattened dome theory was first clearly formulated by the Arab scientist Ibn al-Haytham (c. 1039) in his *Optics*, though it was repeated in very similar terms by the thirteenth-century writers Bacon, Pecham, and Witelo. The part on the moon

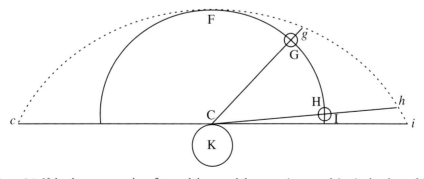

Figure 8.1 If the sky appears to be a flattened dome and the moon is seen as lying in the plane of the sky, then the moon will appear closer and smaller towards the zenith than on the horizon. (From Desaguiliers 1736a.)

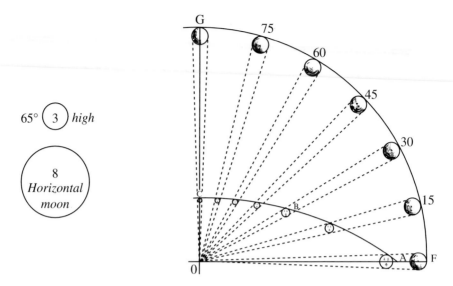

Figure 8.2 Alternative form of the flattened dome according to Smith (1738).

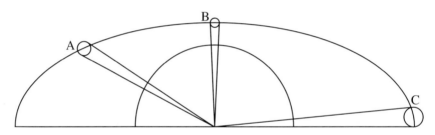

Figure 8.3 Alternative form of the flattened dome according to Young (1807).

illusion consists of several pages of repetitive discourse.[2] The passage can be summarized as follows. The size of an object is judged by combining its visual angle with its known distance. The distance can be properly judged only if the observer can see an uninterrupted sequence of intervening bodies between himself and the object. When looking up at the sky there are no intervening objects – we perceive only an expanse of colour. This appears to us like a blue surface. Because we are familiar with plane surfaces such as walls, the sky is seen as a flat plane extending in all directions. Such a plane must appear closer directly overhead than towards the horizon. The celestial bodies appear to move along this plane and are therefore judged to be further away when on the horizon than when overhead; consequently they are seen as larger on the horizon.

This theory assumes that the perceived size of the celestial bodies is analogous to linear rather than angular size. It is similar to other perceived distance theories of the moon illusion, which will be discussed in Chapter 9. Here we will mainly be concerned with the perceived form of the sky as a phenomenon in its own right.

The sky illusion

The idea that the sky has a definite shape, or even that it is a physical roof covering the world, is found in the mythologies of many countries. In the hero tales of the Yakuts, for example, the sky is described as a hemisphere resting with its rim on the earth. Similarly, the Buriat tribe described its shape as that of a great overturned cauldron.[3] According to a Chinese creation myth of the third century AD, the dome of the sky was shaped from the skull of the dead god P'an Ku.[4] Many tribes in northern Eurasia thought of the sky as a roof to protect the world. Some described it more specifically as a tent roof, often supported on a central pillar. This idea is found among the Lapps, Finns, Yakuts, Japanese,[5] and Hebrews.[6] In ancient Egypt, the most popular conception of the sky was that of a blue river or lake, which was seen as a continuation of the sea and the Nile over which the sun sailed in a boat. A rarer Egyptian view was that the sky consisted of a metal roof or dome. This may have been inspired by observing the fall of metal meteorites.[7]

Similar ideas were entertained about the night sky, which was also described by some north Asiatic tribes as a tent roof. The Milky Way was occasionally seen as a stitched seam, and the stars as tiny holes. The gods would sometimes open the sky for a second or so to observe the earth, during which time a meteor was released.[8] In the mythology of Finland and surrounding countries, on the other hand, the Pole Star was named the 'nail of the sky', on which the celestial dome was thought to be supported and around which it revolved.[9] In the fourteenth-century Welsh nature poetry of Dafydd ap Gwilym the stars are described as 'a set of pieces of shining make for diceing and backgammon, on the vast game-board of the sky'.[10] In Africa, a widespread belief is that the daytime sky is a solid blue vault resting on the earth, on which the sun moves; during the night the sun is thought to return to the east either by passing under the earth (e.g. the Luyia of Kenya) or behind the solid sky (e.g. the Zulus of South Africa), in the latter case producing stars by shining through holes in the vault.[11]

Although these speculations implied a fixed form for the sky, that form was not usually described in detail. The idea that the sky had a flattened appearance did not originate with Ibn al-Haytham, and its origin is difficult to trace. It is ascribed to Empedocles, a fifth-century BC Greek philosopher, by the philosopher Aetius (probably first century AD): 'Empedocles says that where the height of the sky from the earth is concerned, the distance away from us is more than the distance across the width. The sky is more extended there since the world order lies like an egg.'[12]

A clue to its early introduction in astronomical theory may be found in the thirteenth-century writings of John of Sacrobosco, who repeated a proof that the celestial dome is spherical, derived from the ninth-century Arab astronomer al-Farghani (known as Alfraganus to medieval authors):

> Also, as Alfraganus says, if the sky were flat, one part of it would be nearer to us than another, namely that which is directly overhead. So when a star was there, it would be closer to us than when rising or setting. But those things which are closer to us seem larger. So the sun when in mid-sky should look larger than when rising or setting, whereas the opposite is the case; for the sun or another star look bigger in the east or west than in mid-sky.[13]

For al-Farghani the moon illusion was contrary to the effects of a flattened sky, whereas Ibn al-Haytham argued that a flattened sky could explain the illusion. The apparent contradiction illustrates again how careful one should be with the exact meaning of terms. What al-Farghani and John of Sacrobosco claimed was that if the celestial dome had a physical existence and if it were really flattened, then the midday sun would actually be closer, and therefore larger in angular size, than at sunrise and sunset. In contrast, Ibn al-Haytham assumed that the sun was in fact equidistant throughout the day, and had a constant angular size; however, the flattened appearance of the sky makes us judge the sun to be closer at midday, and therefore smaller in perceived linear size than at sunrise and sunset. Thus the contradiction depends on whether one assumes the celestial dome to be really flattened, or only perceived as flattened. In Ibn al-Haytham's time, and even earlier, the idea of an actually flattened dome was untenable; Ptolemy had argued against it, much as al-Farghani did later, when considering whether the setting sun disappears to infinity. Ibn al-Haytham was probably aware of the argument about the angular size of the moon and its relation to the form of the celestial sphere, and by a change of conceptual framework transformed it into a promising explanation of the moon illusion. It is interesting to note that according to Ibn al-Haytham the perceived shape of the sky was a flat plane – similar to the presumed real shape of the heavens against which Ptolemy and al-Farghani directed their arguments.

The flat-sky theory raises several questions. Can the perceived form of the sky be adequately measured? Is it approximately the same for all people? What causes the sky to appear flattened (if indeed it does)? And do the celestial bodies really appear to be at the same distance as the sky? Much has been written on these questions, and we will attempt to summarize only the main points.

Measuring the perceived form of the sky

Attempts to quantify the perceived form of the sky date back to the work of Johannes Treiber in the seventeenth century.[14] Treiber assumed that the perceived surface of the sky results from scattering of sunlight at the top of the earth's atmosphere: the sky therefore forms a curved plane, parallel to the earth's surface, and completely surrounding it. Measurements of the duration of twilight convinced Treiber that the height of the overhead visible sky was about 30 km. Assuming this height, a simple calculation based on Pythagoras' theorem shows that for an observer on the surface of the earth, that part of the sky surface seen next to the horizon is about 20 times as distant as the sky directly overhead. Treiber assumed that this ratio also held for the perceived distances of the horizon and zenith sky, and not just for the actual distances to the scattering layer at the top of the atmosphere. However, he argued that the perceived distance to the sky was uncertain, because there are no intervening objects to act as depth cues (except towards the horizon); the horizon sky therefore appears to be at the same distance as the observer's horizon, whereas the zenith sky appears about one-twentieth as far. The next part of Treiber's argument rests on several arbitrary and unconvincing assumptions.

He argued that, from the usual height of an observer's eye above the ground (not more than about 4 m he says), the earth's surface appears concave, sloping up to the horizon a few kilometres away (compare Fig. 8.10). Treiber assumed that the depth of this illusory terrestrial hollow was the same as the height of the sky vault above the horizon. The sky vault and terrestrial hollow, when combined, therefore cause the horizon sky to appear only ten times as far as the zenith sky, instead of the 20 times implied without the terrestrial hollow. Treiber's work received little recognition from other authors, and it has failed to influence later views on the perceived form of the sky.

The modern era in these investigations began with the publication of an extensive treatise on optics by Robert Smith (1738). His main contribution was to devise a method to measure the ratio of the perceived distances to the horizon and zenith skies. His method was based on the position of the point that was judged to be half-way between the horizon and zenith along the arc of the sky (Fig. 8.4). Once an observer has located the mid-point, its angle of elevation can be measured with a theodolite. The angle is usually found to be less than the required 45 degrees: in fact Smith found it to be only 23 degrees on average. He assumed that the visible sky has a spherical shape, but that the centre of the sphere lies some distance beneath the observer – a suggestion made earlier by the philosopher Thomas Hobbes.[15] Smith calculated the horizon-to-zenith distance ratio to be about 3.3 to 1 (p. 64). He used this ratio, and the assumed spherical shape of the sky, to predict the perceived size of the sun or moon at any angle of elevation.

Smith's work was questioned on several grounds. One of these was his assumption that the perceived shape of sky forms a section of a sphere. Distance ratios can be calculated from any geometrical form – but there is no compelling reason for any particular choice. The simplest forms are a section of a sphere and half an ellipsoid. The latter is formed by rotating half an ellipse, with its major axis from horizon to horizon through

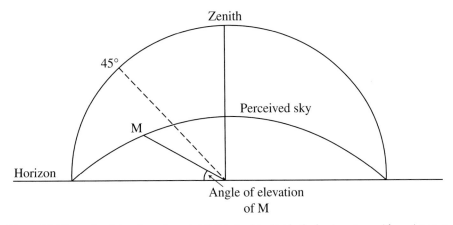

Figure 8.4 The angle of elevation of a point (M) judged to divide the horizon-to-zenith arc into two equal lengths has been used to measure the perceived flattening of the sky. A lower angle of elevation of M implies a flatter sky.

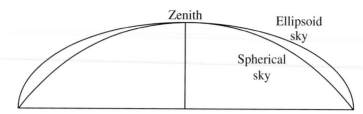

Figure 8.5 Sections through two hypothesized forms of the visible sky: part of a sphere (shown here by part of a circle) and half an ellipsoid (shown here as half an ellipse).

the observer, around its vertical minor axis (Fig. 8.5). Researchers in the biological sciences usually preferred the ellipsoid, while astronomers and physicists usually followed Smith in preferring a sphere.[16] These shapes do not necessarily conform to that of anyone's perceived sky. The different observations and assumptions have been extensively discussed, but without agreement being reached.[17] The differences may be partly due to differences of perceptual set in the observers or, as Angell[18] put it, 'their preconceived opinions of the shape of the heavens or of their manner of viewing them'.

A second question was whether observers agree on the estimated mid-point of the horizon–zenith arc. We shall first consider the daytime sky. The meteorologist Kämtz obtained values of between 19 and 25 degrees,[19] supporting those of Smith. Reimann (1902b) found that his own subjective mid-point had an elevation of between 19 and 22 degrees, varying with the extent of clouding and the season. Miller (1943) found values of about 33 degrees for a clear sky, decreasing to 30 degrees with increasing cloud cover and varying with the nature of the visible horizon. These variable results pose problems similar to those discussed in Chapter 4 for size judgements. What exactly is it that people indicate when they point to the mid-point of the horizon–zenith arc? It could be the point that divides the distance along the perceived arc of the sky into two equal parts, or it could be the direction that divides the horizon–zenith angle into two equal-appearing angles. Reimann (1902b) tested many observers and found that both interpretations were in use, and that some people were uncertain of what was expected of them: he was forced to admit failure in his attempts to determine whether everyone saw the sky as equally curved. Some other observers under the guidance of von Sicherer[20] showed hardly any perceived flattening: 13 people gave estimated mid-points of 38 degrees for a clouded sky, 42 degrees for a clear sky, and 44 degrees during hazy weather. In view of such uncertainties it seems rather pointless to apply elaborate mathematical analyses to this type of result, as was done by several authors.[21]

A more promising method would be to determine the whole perceived form of the sky directly. Robert Smith mentioned a proposal by Martin Folkes[22] to do this. Folkes thought of using a series of fixed points on the ground at various known distances from the observer, who would indicate corresponding points in the sky judged to be directly above each of these (Fig. 8.6). By measuring the angle of elevation of these points in the sky the dimensions of the perceived sky can be quite easily established.[23] This method is so obviously superior to that of Smith that it is difficult to understand why it has never

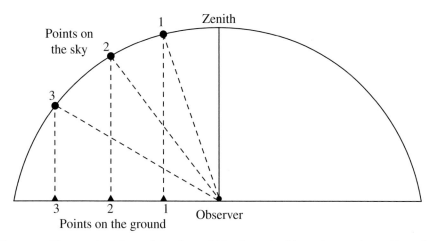

Figure 8.6 The perceived form of the sky could be determined by points on the visible surface of the sky that are judged to be perpendicularly above points on the ground at known distances from the observer.

been used. Quite recently, however, the idea of subjectively corresponding sky and ground points has been incorporated into a perceived distance theory of the moon illusion (Baird 1982) which we will consider in Chapter 9.

The perceived form of the night sky is no more certain than that of the day sky, though several investigators have claimed that it is less flattened. Von Sicherer placed the subjective mid-point of the horizon–zenith arc at an elevation of 42 degrees,[24] and Dember and Uibe (1918) placed it at 40 degrees for a moonless night and 37 degrees with the moon present. More recently the clear night sky was investigated by Baird and Wagner (1982), who obtained estimates from 24 observers of the distance to the sky at 12 different angles of elevation, relative to either the horizon or the zenith sky. Most observers judged the form of the sky as somewhat peaked, but some as slightly flattened. Another 20 observers estimated the distances to various patches of horizon sky that appeared adjacent to buildings at different known distances. The estimates were again very variable, but the sky tended to appear further away when close to more distant objects on the horizon. This latter finding confirms the personal observations of earlier investigators.[25]

Afterimages in the sky

If one looks fixedly at a bright light for a while, an afterimage of the light source is seen as if floating in space directly in front of one's face. Wherever one looks for the next half minute or so, this image is superimposed on the scene. By turning the eyes in different directions, the afterimage can be viewed against different backgrounds. Various authors noted that an afterimage appears to change in size in proportion to the distance (or perceived distance) of the surface onto which it is projected (Fig. 8.7). Benedetto Castelli (1639) was probably the first to describe the effect, though he interpreted it

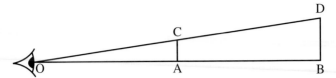

Figure 8.7 The perceived distance interpretation of Emmert's law: the projected image DB appears twice as large as CA because it appears to be located at twice the distance.

as an example of relative size.[26] Emil Emmert (1881) formulated these observations in what came to be known as Emmert's law. There is, however, some controversy as to what Emmert intended. Did he mean the perceived or the actual distance of the projection surface, and did he mean the perceived or the projected linear size of the afterimage? Most authors have assumed that Emmert intended perceived size and distance.[27] An alternative interpretation is that actual distance is the determiner, the projected size of the afterimage increasing with the distance of the projection surface. Thus if the projected sizes of AC and BD in Fig. 8.7 could be measured with a ruler or by means of markers, BD should be twice the size of AC.[28]

If we accept the usual interpretation of Emmert's law, changes in the perceived size of afterimages at different elevations in the sky can be used to map the perceived form of the vault. Zeno (1862) was probably the first to report such changes. He looked at the setting sun to obtain an afterimage, and noted that the image appeared smaller against the zenith sky, but the same size as the sun when viewed against the horizon sky. The magnitude of the size change seems to have been about a factor of 2 (Lewis 1862). Zeno argued that the size change followed from the flattened appearance of the sky, in accordance with Emmert's law. Similar observations were made by several authors in the late nineteenth and early twentieth century.[29] The afterimage illusion was also rediscovered by several later investigators. Trotter (1938) discussed it briefly, while King and Gruber (1962) again considered it to be 'a new phenomenon'. Still later, Wenning (1985) described how he discovered the illusion for himself.

The early observers did not usually quantify the size of the effect, but some later authors included measures that helped to describe the shape of the sky. Angell (1932) found that an afterimage of the sun appeared smaller against a background about 120 m away than against the horizon sky, and that it appeared smaller still against the blue sky at an elevation of about 30 degrees. This seems to indicate that the raised sky was perceived as closer than 120 m, whereas the horizon sky was perceived as further.

King and Gruber (1962) introduced some important measures in their study. Their 16 observers viewed the afterimage of a bright blue paper square, subtending an angle of about 5 degrees, against the horizon sky and at elevations of 45 and 90 degrees. The observers made ratio judgements of the perceived sizes of the image on the horizon and at the two elevations. The results of this simple experiment considerably enriched our knowledge of the illusion. First, the illusion was confirmed by 14 of the 16 observers even though they had one eye covered: for most people, therefore, the illusion does not depend on using both eyes. Second, the experiment yielded quantitative results. For the

14 observers who made one-eyed estimates, the mean ratio of estimated size was 1.50 for the 45 degree comparison, and 1.63 for the zenith comparison. Individual estimates were quite variable. The monocular afterimage illusion is therefore quite substantial, and as large as many measures of the moon illusion. A third result concerns the possible effect of clouds. Eight observers were tested with clear skies, five with wisps of cloud present and three with an overcast sky, but with no obvious differences in their results. Clouds tend to increase the extent to which the sky appears flattened, so the results might suggest that the afterimage illusion does not depend on the perceived form of the sky. However, the number of observers was too small to reach a firm conclusion.

A few years later the illusion was studied along similar lines by Kamman (1967). His observers compared an afterimage viewed against the horizon sky with one viewed at 45 degrees, and gave mean ratio estimates of 1.44 (men) and 1.17 (women). The sex difference is not enlightening. Kamman's most interesting results were obtained by persuading his observers to estimate the relative perceived distance of the afterimage in the two positions – something that Zoth (1899) had found impossible to achieve. The results were clear, but contrary to expectations based on the flat-sky theory and the usual interpretation of Emmert's law: the afterimage appeared further away at an elevation of 45 degrees than on the horizon. We discuss similar distance estimates with regard to the moon and sun in Chapter 9.

The afterimage illusion has almost invariably been attributed to the difference in perceived distance between the horizon and zenith skies. Yet, like the moon illusion, it has also been observed against the night sky (Lewis 1862) which has no clearly perceived distance or form. This does not necessarily mean that perceived distance plays no role in the illusion, but rather that the perceived distance of the afterimage must be determined by factors other than the form of the sky.

The afterimage illusion seems so closely related to the moon illusion that it is tempting to ascribe them to the same cause. Indeed, afterimages have frequently been used to test explanations of the moon illusion, as described in later chapters. If we accept a single underlying cause, a comparison between the factors affecting the two illusions could help to rule out at least some explanations. For example, the size of an afterimage cannot be affected by refraction in the atmosphere; so even if the latter could explain the moon illusion it could not relate to afterimages. Similarly, the atmospheric effects of colour and luminance do not directly affect afterimages – though they could, of course, affect their contrast with the sky background. Afterimages are also unaffected by changes in pupil size, or inadequate focusing of the eye, which we considered in the previous chapter. However, if we accept that the moon illusion can have more than one cause, these other factors cannot be ruled out purely on the grounds that they do not affect afterimages.

Explanations of the sky illusion

If we are to use the flat-sky theory to explain the moon illusion, we must first explain the sky illusion itself. Several attempts, such as those of Treiber and others,[30] have been made to explain the latter as an optical phenomenon. These accounts assume that what we perceive as the surface of the sky is actually closer to us towards the zenith than towards the horizon. The usual argument is that the light reaching the eye from the zenith sky travels a shorter distance than that from the horizon sky, because of the greater thickness of air in a horizontal direction. This type of explanation requires that the observer should be able to perceive such a difference in distance. But of course this is not the case, just as the difference in distance between the sky and the celestial bodies cannot be perceived. The actual distance that light scattered by the atmosphere travels before reaching the eye is immaterial, and optical theories of the form of the sky need not be considered further.[31]

The idea that the sky illusion is a perceptual rather than a physical phenomenon is supported by various observations, though at the time of Bourdon's critical review (1897) no agreed explanation was established. Some authors noted that the perceived shape of the sky varies with how one looks at it: it changes with bodily orientation and the angle of regard, just as has been claimed for the moon illusion (Chapters 11 and 12). For example, Filehne (1894) observed that when his head was pointing downwards the flattening vanished and both the day and night sky appeared hemispherical. Similarly, Zoth (1899) found that when he was lying on his back the sky looked further away towards the zenith than when he was standing up, especially after a few minutes. Minnaert (1954, p. 163) suggested some more dramatic bodily positions: 'Hang by your knees from the horizontal bars and look round you while your head is hanging down. The sky will look hemispherical.' A change with viewing time was also described by Pernter and Exner (1922, p. 5), who noted that the perceived flattening of the daytime sky tends to disappear after staring fixedly at any point.

Other authors have emphasized the role of the visual scene rather than of bodily orientation. Bourdon (1898) noted that the horizon must be visible for the sky to appear flattened, and that in misty weather it looks hemispherical. Intervening objects may therefore play a part, as was suggested by Ibn al-Haytham. Humphreys (1964, p. 453) suggested yet another factor, aerial perspective: the overhead sky is clearer than the hazy horizon sky, and experience makes us see clear objects as nearer. Many authors have stressed the importance of perceptual learning from the flat or vaulted appearance of a cloud-covered sky (Fig. 8.9). Robert Smith[32] described this appearance, and suggested that the same idea might apply to a clear sky:

> And when the sky is either partly overcast or perfectly free from clouds, it is a matter of fact we retain much the same idea of its concavity as when it was quite overcast. But if anyone thinks that the reflection of light from the pure air, is alone sufficient to suggest that idea, I will not dispute it.

Smith therefore hedged his bets, by suggesting that there might be an optical explanation. The perceptual learning account was rejected by J. C. E. Schmidt in 1834 on the grounds that the (real) flattening of a clouded sky was much too pronounced to explain the slight perceived flattening of clear skies.[33] However, this is not a serious objection, because the perceived shape of the clear sky may be a compromise between the remembered strongly flattened cloudy sky and the presently visible formless expanse of blue.

Helmholtz[34] supported the idea of perceptual learning from clouds, an explanation that is still attractive to current writers: for example, Walker (1978b) assumed that the flattened appearance of the daytime sky carries over into our perception of the night sky. Nevertheless, there is little evidence that the night sky appears flattened – a fact that causes difficulties for all such explanations.

Celestial bodies and the flattened sky

Our final question relating to the flat-sky theory is whether the perceived distance of the celestial bodies is the same as that of the sky – something that is essential if one wishes to use the sky illusion as an explanation for the moon illusion. There is some agreement among observers that the two coincide for the daytime sky.[35] However, a large-scale investigation by Zoth (1899), in which he questioned about a hundred people, did not confirm it for the night sky. Zoth asked his subjects whether they could remember a particular moonrise when the moon appeared very large. The answer was almost invariably yes. He then asked whether on this occasion the moon had appeared to be located on the surface of the sky or had seemed to float in space in front of it. The answer was usually the latter. The stars, too, may appear separated from the sky surface: Indow (1968) reported that his subjects had the impression that a star seemed to recede into the sky when fixated.

We are forced to conclude that the sky illusion does not provide an adequate explanation of the moon illusion. In the first place, the daytime sky does not appear like a flattened dome to everyone, and its perceived form does not usually match the magnitude of the moon illusion at various angles of elevation. Second, the night sky is not usually seen as a flattened surface, though the celestial illusion is often quite pronounced after dark. Third, the enlarged horizon moon is not usually seen as if fixed to the night sky, but seems to float in space in front of it. Finally, the sky illusion is itself a perceptual phenomenon in need of further explanation and is therefore hardly a sound basis on which to explain the moon illusion. Further attempts to explain both the sky illusion and the moon illusion in terms of perceptual learning will be considered in Chapter 9.

How high is the sky?

Previously we considered the perceived form (or relative dimensions) of the sky vault, rather than its perceived absolute distance. The perceived distance may also be important, if it is to be used to explain the celestial illusion. Unfortunately it is difficult to obtain any truly independent measures of the perceived distance of the sky vault separately from that of the celestial bodies. The distance of the sky vault is usually judged indirectly, through estimates of the perceived distance of objects that are assumed to lie on its surface.

The absolute perceived distance of the celestial bodies is very indefinite, which may have discouraged investigators from collecting estimates. The relative perceived distance of these bodies on the horizon and in the zenith is an easier judgement, and we discuss such reports in Chapter 9. The only absolute estimate of which we are aware was reported by Bourdon (1897), who claimed that most people see the stars as about 80 to 150 m distant. However, much smaller values are obtained if indirect methods are used to calculate the 'registered' distance of the celestial bodies. The registered distance is a hypothetical rather than a perceptually estimated distance: it can be calculated from estimates of the perceived linear sizes of the celestial bodies, combined with their known angular sizes. Von Sterneck (1907, 1908) used the method to obtain a value of about 12 m for the elevated celestial bodies. Common estimates of the absolute size of the sun and moon lie between 100 to 300 mm (Chapter 3); and for an angular size of about half a degree, such estimates could be taken to imply a perceived distance of 12 to 36 m. A study of various units that were used by ancient astronomers in different cultures to measure both length and angles yields very much the same result.[36] It is not clear how these very close distances should be interpreted, as they do not correspond to the directly estimated distance and do not take into account any possible misperception of angular size. They merely represents the true distance at which the moon would have to be for its estimated linear size to subtend its true angular diameter. We mentioned the difficulties of interpreting linear size estimates in Chapter 3, and we shall return to the concept of a registered distance in Chapter 9.

A different method for calculating the perceived distance of the sky involves the use of afterimages, which we described earlier in this chapter. They can be projected to different distances on the ground, and their perceived size then compared with that of afterimages projected in the sky. Plateau (1880) used this method. He formed an afterimage of the full moon on a clear night and viewed it against a nearby wall; he then changed his distance from the wall until the image appeared the same size as the moon had done a moment earlier. At this point his distance from the wall was 51 m, which is therefore an indirect measure of the moon's perceived distance. Other observers are said to have confirmed this, reporting registered distances of 50 to 60 m.[37] We have previously noted the similar finding of Angell (1932) that the zenith sky was closer than 120 m.

Von Sterneck's quantitative distance account

Von Sterneck (1907, 1908) developed a quantitative theory of distance perception, which could account for the flattened dome and the moon illusion. He noted that, in general, perceived distance increases much more slowly than true distance. As a result distant objects all appear equally far. The maximum perceivable distance depends on the amount of illumination and other conditions, being around 200 m at night and 20 km by day. Von Sterneck expressed the perceived distance of an object (d') in terms of its true distance (d) and the maximum perceivable distance under the particular circumstances (c) by the following expression:

$$d' = cd/(c+d).$$

According to this formula, the perceived and true distances are practically equal when the maximum perceivable distance is well beyond the true distance, but underestimation

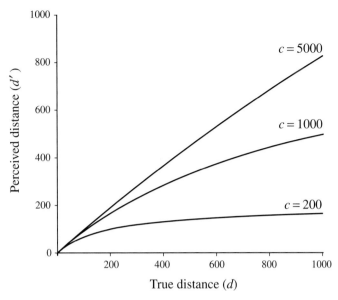

Figure 8.8 The relation between perceived distance (d') and true distance (d) according to Von Sterneck (1907, 1908). The greater the maximum perceived distance (c), the more closely d' approaches d.

increases when the two are close together. Examples are shown in Fig. 8.8 for c equal to 200, 1000, and 5000.

Von Sterneck used this formula to account for the vaulted appearance of the clouded sky. A flat cloud bank overhead will not be seen as flat but rather as sloping down to the horizon, because the distance to the further clouds is underestimated more than that to the nearer clouds overhead. This idea is illustrated in Fig. 8.9. If distance perception and size–distance invariance were perfect, the cloud bank would be seen as flat with clouds of equal size. However, if no distance cues were available all the clouds would be seen as equidistant, and the bank would appear like a perfect hemisphere with the cloud size diminishing towards the horizon. The curved appearance of the clouds could therefore be accounted for by foreshortened distance perception.

If the true and perceived distances of some of the clouds are known, von Sterneck's formula can be used to calculate first the maximum perceivable distance, and then the perceived shape of the whole vault. For example, a cloud bank may be at a height of 2.5 km at the zenith, and visible at 177 km on the horizon. If an observer estimates that the horizon clouds appear to be five times as far as the clouds overhead, some algebraic manipulation indicates that the maximum perceivable distance is 10.6 km. This value can then be used to calculate the size and shape of the whole sky vault under the particular conditions. Von Sterneck used the same formula, with different values of c, to describe the flattened vault of the clear blue sky and the night sky. He then assumed that the flattened vault formed a reference surface on which the celestial bodies appeared to lie, and that their perceived linear sizes would follow from size–distance

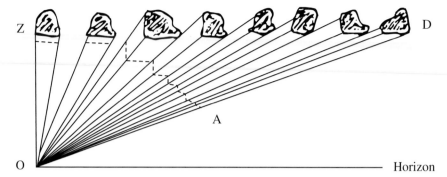

Figure 8.9 Perspective view of the clouds. An observer at O sees a horizontal layer of clouds (ZD) as a stair-step pattern (ZA), the further clouds appearing as near as the zenith clouds. (After Humphreys 1964, p. 451. From Ross 1974, with permission of the publisher.)

invariance. His reference level theory of the celestial illusion therefore amounts to a quantified form of the sky illusion explanation.

A major weakness of this theory – as far as explaining the celestial illusion is concerned – is that the form of the reference surface for the celestial bodies cannot be determined independently. Von Sterneck derived it from the estimated linear sizes of the sun and moon at various angles of elevation, obtaining distances similar to those discussed above. Using these in turn to explain the moon illusion constitutes a circular argument that does not advance our understanding at all.

Von Sterneck's theory was criticized on similar grounds by Müller (1907). Müller (1921) also maintained that it was not legitimate to write an equation relating perceived to physical distance, as one term was psychological and the other physical. This point went unheeded, however, as psychophysicists had been relating perceptual to physical measures since the writings of Fechner (1860). Modern psychologists have continued to quantify perceived distance, and several have confirmed that estimated distance increases more slowly than physical distance, at least in natural outdoor settings.[38] Minnaert (1954) agreed that distance underestimation can explain many terrestrial phenomena,[39] but he criticized von Sterneck's attempt to measure the perceived distance of the empty sky.

The terrestrial saucer

We have seen how the 'hollow earth' entered into Treiber's account of the celestial vault, but the phenomenon deserves a little more attention. The concave or hollow earth, as seen from on high (Fig. 8.10), is the converse effect of the arching clouds seen from the ground. From the viewpoint of a camera in mid-air, one is surrounded by a sphere: in the absence of any contrary information, all points on the earth's surface can be considered equidistant. There is, of course, information present in the form of texture gradients and other cues indicating that the horizon is further away than points directly below. The fact that the earth's surface appears concave rather than flat indicates

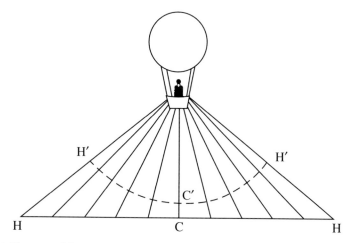

Figure 8.10 The terrestrial saucer or concave earth. Failure to appreciate that H is further away than C makes the flat surface HCH become the curved surface H'C'H'. (From Ross 1974, with permission of the publisher.)

that distance perception is imperfect, the horizon distance being underestimated in comparison with the vertical distance. Thus the arching clouds and the terrestrial saucer are both examples of foreshortened distance perception.

If the celestial bodies could be viewed as moving on the surface of the terrestrial saucer, they should change in perceived size in a similar manner to their changes in the sky. They should appear at their largest on the horizon, and become smaller as they sink down into the hollow of the ground. We cannot, of course, view the celestial bodies in this manner, but there are many reports that ordinary objects appear much smaller when viewed from a height than when viewed at a similar distance horizontally.[40] We can, however, view afterimages or artificial moons projected downwards. Such experiments will be described in Chapter 12.

Heelan's hyperbolic visual space

The perceived sizes and distances of the celestial bodies, the terrestrial saucer, and many other visual phenomena are covered by a quantitative theory of visual space proposed by Heelan (1983). Like von Sterneck and others, Heelan notes that we usually experience our close physical environment as an undistorted three-dimensional space (a so-called *infinite Euclidean space*). However, our perception of distant objects, visual illusions, and various other visual experiences imply that visual space (as opposed to physical space) can often best be described as a so-called *finite hyperbolic space*. In such a space, for example, distant objects seem smaller and have less depth than closer objects, distant parallel lines appear to converge, and horizontal planes – such as a clouded sky or the earth's surface seen from on high – appear to curve round the observer.

Heelan has described such hyperbolic spaces mathematically. The visual cues available in a given situation determine the shape of an observer's hyperbolic visual space. Thus the characteristics of the visual space when viewing the moon on the horizon are different from those when viewing the elevated moon, although both visual spaces are finite and hyperbolic. Heelan uses the fact that the horizon moon appears both larger (in linear size) and closer than the elevated moon to determine the differences between these two visual spaces; and in this way he attempts to provide a description of the illusion which links it to other perceptual phenomena, including the terrestrial saucer and sky vault.

An important consequence of Heelan's theory of the moon illusion is that horizon objects in the vicinity of the rising moon should also appear enlarged. Heelan claims that this is indeed what we experience: distant trees and houses appear larger when seen close to the horizon moon than when the moon is elevated. A phenomenon of this type was described by Dunn (1762) and Henning (1919), but only in connection with the red light of the rising or setting sun: both of their accounts are quoted in Chapter 6 in the section on changes in colour. There are no other reports of changes in the scale of terrestrial objects with the varying elevation of the sun or moon, so an unsubstantiated phenomenon cannot be used to provide support for Heelan's theory.

The main problem with Heelan's explanation, however, is that, like von Sterneck's theory, it is circular: the appearance of the horizon moon is used to derive the nature of the observer's visual space, which is in turn used to explain the moon illusion; the parameters of visual space are not determined independently.

We have to conclude that no authors have succeeded in establishing the perceived form of the sky as an independent explanation of the moon illusion. There are, however, several similarities between the two illusions, so they probably share some causal factors.[41]

Summary

The sky is reported to appear as a flattened dome. Ibn al-Haytham and others argued that the celestial bodies appear to follow the path of this dome, and that the change in perceived distance accounts for the change in perceived (linear) size through size–distance invariance. However, attempts to measure the perceived form of the sky produce variable results; the night sky generally appears less flat than the day sky; and observers report that the moon appears to float in front of the sky. The perceived size of afterimages in the sky varies in the same way as the celestial illusion, but their perceived distance does not coincide with the flattened dome. Estimates of the perceived distance of the sky and of the perceived linear sizes of the sun and moon are inconsistent with classical size–distance invariance, because calculations give very close 'registered' distances. The quantitative distance theories of von Sterneck and Heelan calculate the perceived form of the sky from the estimated linear size of the celestial bodies, and so cannot be used to explain the latter. The sky illusion, and the perceived size of afterimages in the sky, probably share some causal factors with the celestial illusion.

CHAPTER 9

So near and yet so far

One of the most prevalent beliefs about size constancy is that it is achieved by taking distance into account. This belief is often expressed in terms of classical size–distance invariance: for an object with a given angular size, its perceived linear size increases in proportion to its perceived distance. The scale of such changes would be more than sufficient to account for the moon illusion. In support of this explanation, it is claimed that the celestial bodies appear to be further away on the horizon than in the zenith. This idea is closely linked to that of the apparently flattened dome of the sky, which we discussed in the last chapter. We concluded that the sky illusion could not be used as an independent explanation of the moon illusion, partly because the moon does not usually appear to lie on the dome of the sky. We now consider the evidence concerning the perceived distance of the celestial bodies. Perceived distance is a more controversial topic than perceived size: everyone agrees that the horizon moon appears larger than the raised moon, but observers are less certain about its relative perceived distance.

The perceived distance of the celestial bodies

Many observers have claimed that the perceived distance of the rising moon or sun is the same as that of the horizon over which it is seen. This was the view of the Roman poet Lucretius (*c.* 98–55 BC):

> And again, when nature begins to raise on high the sunbeam ruddy with twinkling fires, and to lift it above the mountains, those mountains above which the sun seems to you to stand, as he touches them with his own fire, all aglow close at hand, are scarce distant from us two thousand flights of an arrow, nay often scarce five hundred casts of a javelin: but between them and the sun lie vast levels of ocean, strewn beneath the wide coasts of heaven, and many thousands of lands are set between, which diverse races inhabit, and tribes of wild beasts.[1]

Simply stated, Lucretius said that the rising sun appears to be as far as the horizon, while in fact it is much further. This is, of course, also assumed by those authors who use classical size–distance invariance to explain the perceived size of the celestial bodies. Ptolemy (second century AD) in his late work the *Planetary hypotheses* (Book I, Part 2, Section 7) argued that the celestial bodies seem closer than their true distances and nearer to familiar distances, and their perceived linear size varies proportionately:

> The planets seem closer to us than they really are, for the eye [naturally] compares them to
> things at more familiar distances, as we have explained. The [estimated] magnitude varies
> according to the distance, but at a smaller ratio [than geometric rules would require]
> on account of the weakness of our visual perception to discern quantity of either kind
> [i.e. distance or magnitude].[2]

In this passage Ptolemy did not comment on the different perceived size or distance of celestial bodies in different parts of the sky, so did not explicitly use size–distance invariance to explain the celestial illusion.[3] We discuss Ptolemy's more specific explanations in Chapters 5 and 11.

Some authors have disputed that the moon appears very distant on the horizon. Molyneux (1687, p. 321) wrote:

> my-thinks the *Horizontal* moon should be fancyed nigher to us, than farther from us; for
> if we are for trying Natural thoughts, let us take Children to determine the Matter, who
> are apt to think, that could they go to the edg of that space that bounds their Sight, they
> should be able (as they call it) to touch the Sky; and consequently the Moon seems then
> rather nigher to us than farther from us.

Molyneux seems to be saying that the moon appears to be closer to the ground on the horizon than it does in the zenith: it is thus not entirely clear whether he thought it appeared nearer to an observer who is *not* located at the horizon.

Most authors, such as Dunn (1762, p. 471), explicitly stated that the moon appeared nearer on the horizon. In 1775, G. S. Kügel reported that he had asked several people, and they all claimed that it appeared closer.[4] Similar findings were reported by many authors around the early 1900s.[5] King and Hayes (1966) later claimed that most people remembered the horizon sun as both larger and closer. Recently the archaeologist Ruggles assumed that to be the case (1999, p. 154).

Most of these authors pointed out that these observations created a problem for the size–distance invariance explanation. Boring (1943, p. 56) wrote:

> Yet the [perceived distance] argument is wrong for the simple reason that the moon in
> elevation looks farther away than the moon on the horizon – to all who have observed the
> moon much and thought about the matter. If asked why, they say, 'The moon is so much
> smaller in elevation; of course, because it looks farther away' – thus inverting Ptolemy's
> logic.

All of these reports are based on memory, rather than experiment. Indeed, one cannot compare the perceived distances of the horizon and zenith moon at the same time. It is, however, possible to experiment with afterimages, or with artificial moons. Zoth (1899) found it impossible to determine the perceived distance of afterimages in the sky, but Kamann (1967) found that they were judged both larger and closer when projected onto the horizon; he concluded that Emmert's law did not apply and size–distance invariance could not explain the illusion. Many other experiments with afterimages and artificial moons are described in Chapters 11 and 12, most of which confirm that objects of the same angular size appear both larger and nearer when viewed horizontally.[6]

Most of the reports mentioned in this section concern the relative rather than the absolute perceived distance of the celestial bodies. Estimates of absolute distance were described in Chapter 8, and they suggest that the celestial bodies appear very close when near the horizon. We have to conclude that whereas the horizon itself generally appears further than the zenith, the celestial bodies appear nearer on the horizon. Nevertheless, much effort has been expended on explaining the supposed greater perceived distance of the horizon moon. The most popular explanation involves the role of intervening objects.

Intervening objects

The earliest explanation for the flattened dome was that the intervening objects on the surface of the earth made the horizon appear further than the zenith. Filled space appears greater than empty space, as Plotinus noted in the third century (Chapter 3). The first author to relate this to the flattened dome appears to have been Ibn al-Haytham, who wrote:

> Human vision can never perceive the size of visible objects, unless the intervening space forms a straight line of adjacent and uninterrupted bodies … If vision has not accurately established the distance of the object seen, it is able to estimate its distance and assimilate this to the distances of visible objects with which it is accustomed … The intervening space between stars and observers does not form a straight line of adjacent and uninterrupted bodies … accordingly vision estimates the distance of the stars by comparing it with the distance of those terrestrial objects which are perceived from the greatest possible distance.[7]

This account was repeated, with slight rephrasing, by Witelo, Bacon (*c.* 1263), and Pecham (*c.* 1274). They were following Ibn al-Haytham and made little, if any, contribution of their own. Witelo in one place ascribed the moon illusion to atmospheric vapours but elsewhere to intervening objects:

> because of the breadth of the space of the earth's surface, which is perceived between the viewer and the horizon, while nothing is perceived between the vault's zenith and the earth. Because the extent of the remoteness is judged from the apparent distance of the intervening objects, it follows that the distance is judged greater when a greater apparent quantity seems to intervene. Therefore the distance to the edge of the horizon seems much greater than the distance to the zenith of the apparent vault.[8]

Roger Bacon (*c.* 1263) gave a similar account in his *Opus majus*,[9] presumably without knowledge of Witelo:

> The distance of the stars on the horizon is gauged through the interposition of the earth; but they cannot be gauged in this way when they are in the middle of the sky because the air is not perceptible. Therefore since their remoteness seems greater on the horizon than in the middle of the sky, it follows that they appear to be further away than when in mid sky. Therefore (as before) they appear larger.[10]

John Pecham (*c.* 1274) gave the same explanation in his *Perspectiva communis*. He used intervening objects to explain the apparent distance of the clouds (Proposition 63), and the shape of the flattened dome (Proposition 65, p. 143):

> Proposition 65. The horizon appears further away than any other part of the hemisphere.

> This is evident from Proposition 63, for if distance is gathered from the size of [the inter-vening] bodies, then the distance must appear greater where greater magnitude is seen to intervene. But between the horizon and the observer, the whole breadth of the earth is seen to intervene; between the observer and the zenith, nothing. Therefore the horizon appears incomparably farther away than any other part of the sky.

Pecham may have been following Bacon and Witelo, though he does not cite them directly. The intervening objects explanation has been repeated by a very large number of authors, too numerous to mention individually.[11]

Intervening objects and hypothesis testing

During the seventeenth century authors began to speculate about the outcome of 'tube' experiments: what would happen to the perceived size of the horizon sun or moon if the surrounding objects were blocked from view? The idea for a tube experiment may have come from a statement about atmospheric effects by the Greek astronomer Cleomedes, around the third century AD: 'It is also said that where it is possible to view the sun from deep wells, its appearance is much larger since it is seen through the humid air of the well.'[12]

Riccioli's contribution (1651) is of interest, because he seems to have been one of the first to suggest an experiment. He disbelieved in the importance of intervening objects. He noted:

> If you observe the rising or setting sun in your bedroom or garden in such a way that a hedge or wall or lower edge of the window prevents all view of the intermediate space up to the horizon, and that you see nothing else but the sun, you will however see it as immensely greater than when far from the horizon.[13]

Other authors suggested looking through a tube to remove the effect of intervening or surrounding objects. The contributions of Molyneux (1687) and Goüye (1700) were mentioned in Chapter 7: Molyneux asserted that the illusion remained, Goüye that it disappeared. Malebranche (1693) also had something to say on the subject, though he seems to have measured the effects of a reduction in luminance (Chapter 6) in addition to a loss of terrain. If one looks at the sun through a flat piece of glass, darkened by the smoke of a candle flame so that only the sun is visible and not any intervening terrain, then:

> If the sun is at the horizon the interposition of the glass will make it appear about twice as near and about four times as small [as without the glass]. But if it is high above the horizon, the glass will produce no considerable change either in its distance or in its apparent size.[14]

Berkeley also discussed the matter in his *A new theory of vision* (1709, Section 77). He stated that if a wall were to block off the horizon and the intervening objects, the moon would still appear enlarged, due to its faint appearance and other causes. The same claim was made by Le Cat in 1744, presumably following these earlier authors: 'If one looks at the Moon on the horizon over a wall, through a tube of paper or glass, one no longer sees these mountains, these valleys, etc., which indicate its distance, and yet one sees it [the Moon] as bigger.'[15] Like Berkeley, he subscribed to the aerial perspective explanation. Euler made a similar claim in 1762: 'On looking at the moon in the horizon through a small aperture made in any body which shall conceal the intermediate objects, she nevertheless still seems greater.'[16]

It is not clear how many of these authors actually performed a tube experiment, as opposed to some other sort of experiment, or no experiment at all. The astronomer Biot (1810) may have done so. Biot wrote:

> These illusions cease as soon as one no longer sees foreign objects. One can block out these objects by looking at the moon through a tube, or through a roll of black cardboard, which will allow one to see only the Moon, the opening being completely covered by the disc [of the Moon]. By having the same size opening, the moon seems no bigger on the horizon than at its zenith. It is the same if one looks at the Moon through a smoked glass, because the darkness of its colour permits one to see only the illuminated object, and hides all the rest.[17]

The outcome of a genuine tube experiment is important for theories of the illusion. If the illusion is eliminated when the horizon moon is viewed through a tube, the enlargement must be due to intervening or surrounding objects – that is, to perceived distance or to relative size, or to luminance or colour contrast. If the illusion remains at full strength, it must be due to one or more other factors. If it occurs, but at reduced strength, it must be due to a combination of factors that includes the visual surrounds. Despite the importance of the experiment, there was no agreement among early authors on its outcome.

The most likely reason for the disagreement is that the experiment was performed inadequately, if at all. The size of the reduction tube is critical, as even a small area of terrain may be sufficient to produce an illusion. The categories of answer allowed are also important, since the observer may be forced to classify a small illusion as either a full illusion or as no illusion. The outcome may well depend on the theories of the authors. Several authors' theories are listed in Table 9.1, categorized according to their observations (or beliefs) as to whether reduced viewing conditions eliminated the illusion.

The effects reported are, of course, generally in line with the authors' main explanation for the illusion. On the whole, those observations that are apparently genuine suggest that the illusion is eliminated or much reduced when the horizon moon is viewed through a reduction tube. However, it is difficult to draw any firm conclusions because these observations were not true experiments by modern standards: they were usually made by the author alone, or by very few subjects, and the judgements were of the all-or-none type. In good modern experiments large numbers of unbiased subjects give quantitative measures of their perceptions, and authors perform statistical tests of their hypotheses. Yet hypothesis testing did not become common in psychology until about 1940–45.[18]

Many authors continued to make naturalistic observations, and most have supported the importance of the visual surroundings. Lohmann (1920, p. 98) performed a simple

Table 9.1

Author		Theory	Method
(a) Illusion eliminated			
Castelli	1639	Intervening objects/ relative size	View over hat brim
Malebranche	1693	Intervening objects	Dark glass
Goüye	1700	Divided space	Tube
Porterfield	1759	Intervening objects	Tube
Biot	1810	Intervening objects and aerial perspective	Tube
Brandes	1827[1]	Intervening objects, flat sky, and aerial perspective	Perspective tube
Reimann	1902a	Flattened sky	Tube or filter
Mayr	1904	Vertical small, reduction conditions	Tube or filter
Filehne	1910a, 1917	Flat sky, horizontal experience	Tube
(b) Illusion not eliminated			
Riccioli	1651	Refraction	View over wall
Molyneux	1687	Undecided	Tube
Berkeley	1709	Posture/aerial perspective	Wall
Le Cat	1744	Aerial perspective	Wall or tube
Euler	1762	Aerial perspective	Small aperture
Filehne	1894	Flattened sky	Horizon hidden
Zoth	1899	Angle of regard	Dark glass
Filehne	1910b	Flat sky, horizontal experience	Dark glass

[1]In Reimann (1902a), p. 12.

experiment to prove the importance of the proximity of the moon to the visible horizon. He first observed the enlarged rising moon from the bottom of a hill, standing with his back to the hill. When it had risen some distance above the horizon, the illusion disappeared. Lohmann then climbed over the top of the hill to view the moon from the other side, rising above a new and higher horizon – whereupon the illusion reappeared. Similar observations were later made by Goldstein (1962), who took the results to support the perceived distance explanation.

Several modern authors – most of whom favoured perceived distance – have conducted experiments on the effect of terrain and intervening objects. Rock and Kaufman (1962) found the normal illusion to occur when an artificial moon was viewed in conjunction with terrain, but not to occur when the terrain was hidden from view; and the illusion was reversed when mirrors were used to show the raised moon in conjunction with terrain, and the horizon moon with a sky surround. The presence of the terrain produced a perceived enlargement of about 34 per cent. The authors also found that a distant horizon, showing more terrain, gave a larger illusion (51 per cent) than a near horizon (36 per cent). Similarly, a cloudy sky gave a larger illusion (52 per cent) than a clear sky (34 per cent) – perhaps because the clouds increase the perceived distance to the horizon. Rock and Kaufman also inverted the view of the terrain (by means of prisms) and found that the illusion was reduced (28 per cent compared to 66 per cent) – and again argued that this was because the perception of distance is reduced for inverted scenes. Later, Coren (1992) reported a reduction in the pictorial moon illusion when the pictures were inverted, and gave a similar explanation. The presence of a horizon line was found to be sufficient to give an illusion of 26 per cent by Gruber, King, and Link (1963), who used glowing ping-pong balls at elevations of 2 degrees and 17 degrees in a darkened room. The presence of terrain was shown by McNulty and St Claire-Smith (1964) to increase the perceived distance of an artificial sky background, but the authors did not measure the effect of terrain on the perceived size or distance of the moon; they nevertheless took the finding to support the perceived distance explanation of the moon illusion. A different explanation was given by Iavecchia, Iavecchia, and Roscoe (1983). They performed an experiment similar to that of Rock and Kaufman, and found that the presence of terrain caused an artificial moon illusion of up to 26 per cent: yet they attributed the cause to changes in accommodation rather than to changes in perceived distance (Chapter 7).

We can conclude that intervening objects do increase the perceived size of the moon, and the perceived distance of the horizon, but not that the moon illusion is necessarily caused by an increase in its perceived distance. The intervening objects could directly affect the perceived size of the moon, independently of a change in perceived distance, through cues of relative size (Chapter 10).

Perceptual learning: inexperience of vertical viewing

There is a different class of explanation, which assumes that the perception of distance depends not so much on the presence or absence of certain distance cues, but on our ability to interpret those cues. One variant of this explanation assumes that we see objects as near unless we learn to see them as far. Most of our experience is of horizontal scenes, viewed from an upright bodily posture. We learn to perceive horizontal distances correctly, but this learning does not transfer to vertical scenes or unusual bodily postures (see Chapters 10–12). Thus objects appear too near when viewed upwards from the ground, or downwards from a height. This explanation could account for the flattened dome, and for a reduction in the perceived linear size of objects according to the size–distance invariance principle. Some authors stress the role of experience in horizontal viewing, while others stress that of inexperience in vertical viewing. These could be regarded as two distinct explanations, but since the distinction is often blurred we shall consider them under the same heading.

It is just possible that Ptolemy held a view of this type, as expressed in the *Optics* (see Chapter 11). He may have thought that vertical distances were foreshortened through inexperience. However, no clear statements about this seem to have been made until the seventeenth century, when Wallis (1687, pp. 325–6) discussed the role of experience in horizontal judgements. He emphasized the importance of intervening objects in increasing the perceived distance of the horizon, but added that memory could have the same effect:

> When the Sun or Moon is near the Horizon, there is a prospect of Hills, and Vallies, and Plaines, and Woods, and Rivers, and variety of Fields, and Inclosures, between it and us; which present to our Imagination a great Distance capable of receiving all of these. Or, if it so chance that (in some Position) these Intermediates are not actually seen: Yet having been accustomed to see them, the Memory suggests to us a view as large as is the visible Horizon. But when the Sun or Moon is in a higher Position; we see nothing between us and them (unless perhaps some clouds) and therefore nothing to present to our Imagination so great a distance as the other is. And therefore, though both be seen under the same Angle, they do not appear (to the Imagination) of the same bigness, because not both fansied at the same Distances: But that near the Horizon is judged bigger (because supposed farther off) than the same when at a greater Altitude.

Versions of the foreshortened vertical distance explanation, based on vertical inexperience or horizontal experience, have been repeated by many authors.[19]

Perceptual learning: experience of vertical viewing

The vertical *inexperience* account is unsatisfactory because it does not provide a convincing reason why inexperience should make objects appear nearer rather than further. Indeed, some authors such as Allander (1901) have claimed the opposite. An alternative account based on *experience* of vertical viewing is more promising. Objects viewed overhead or at our feet are normally nearer than those viewed horizontally: we look up at ceilings and clouds and down at the ground, whereas we look straight ahead to the horizon. The scene is thus typically one in which the angular sizes of objects are relatively large overhead and underfoot, but small when viewed horizontally. To compensate it is necessary to contract the perceived size and distance of objects viewed vertically in comparison with those viewed horizontally. Thus, experience of vertical viewing causes the elevated moon to appear smaller and the sky dome to appear flatter.

Ibn al-Haytham appears to have been the first to mention perceptual learning as the cause of the flattened sky dome. He proposed it as a cause in addition to the presence or absence of intervening objects:

> However, the full extent of the sky is not apparent to the senses ... and all that vision can sense of the sky is its bright colour ... When ... vision has perceived some colour in its length and breadth, it also perceives a shape and form, and it will then perceive the plane itself: for it will assimilate this [plane] to some everyday surface area, e.g. a wall etc. ... Therefore vision perceives the surface of the sky as flat.[20]

Huyghens (1629–95) believed that perceptual learning about clouds and intervening objects could explain the celestial illusion:

> But to point out the cause of the error [the sun illusion] in a few words, it must be understood that this is the origin: we judge the Sun or any other heavenly body near the horizon

to be farther away from our eye than when the same body approaches the vertex. This is because we imagine things overhead in the sky, when they are far from the horizon, to be no farther away from us than the clouds that float above our heads. On the other hand, we are used to observing a large breadth of ground intervening between us and things near the horizon, with the arch of the sky appearing to begin at the end of the ground. Therefore we are accustomed to judge both it, and objects visible in it, as far away from us. So now when two bodies of equal magnitude cover the same visual angle, we always judge as larger the one that we take to be farther away.[21]

Our experience with clouds and overhead objects was thought by many authors to be important in explaining the illusion.[22] A version of the theory was given more elaborate form by Reed (1984, 1989, 1996) who called it his 'terrestrial passage' theory; but since it is an angular size theory rather than a distance theory it is described in the next chapter.

Environmental experience and spatial perception

Explanations of the moon illusion based on the effect of experience on perceived distance are difficult to substantiate. One would need to compare the size and distance judgements of different groups of people who were accustomed to see either near or far distances in either the horizontal or the vertical plane. Such groups are hard to find, since most people live in environments in which overhead and underfoot objects are closer than objects near the horizon.

The only evidence appears to be anecdotal, or to relate to one or two individuals. For example, Koffka (1936, p. 279) cites a study by von Allesch (1931) who examined the size and distance discrimination of humans and of one lemur – an animal accustomed to leap up and down trees: human judgements were finer when viewing in the horizontal than the vertical orientation, but the reverse was true of the lemur. Koffka noted that the experiment did not directly compare perceived size in the two orientations, and also commented: 'Perhaps one such experiment is not quite sufficient to prove so radical an assumption … It is to be hoped that new experiments will decide this highly important issue.' Koffka's hopes were not fulfilled. Leibowitz and Hartman (1960a) mentioned that two adults showed no moon illusion in an experiment, one being a forest ranger and the other an amateur pilot, and suggested that their vertical experience may have reduced the illusion. In a similar vein, Turnbull (1961) reported that a young pygmy man from a Congo forest was unable to appreciate the size or distance of buffaloes when he was taken for a drive in open country: he thought they were insects which grew in size as he approached them. His lack of size constancy may have been due to lack of experience of large horizontal distances.

Whereas little is known about the effects of environmental experience on the moon illusion or size constancy, many investigations have been conducted concerning the geometrical illusions. There is some evidence that people living in towns tend to be more susceptible to the Müller–Lyer illusion (the arrowhead length illusion – Fig. 9.1) than those living in flat open environments, while the opposite is true for the horizontal–vertical illusion (the lengthened verticals illusion – Fig. 9.2). It is open to question

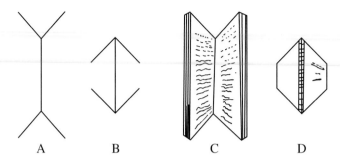

Figure 9.1 The Müller–Lyer illusion. The vertical line in A appears longer than in B. This might be because it is taken as representing a relatively distant object such as the interior spine of an open book, whereas the line in B is taken as representing the relatively near exterior spine of a book. The typically far line is then perceptually enlarged in comparison with the typically near line. Experience of built environments might bias observers towards this interpretation of the figures. (From Ross 1974, p. 60, with permission of the publisher.)

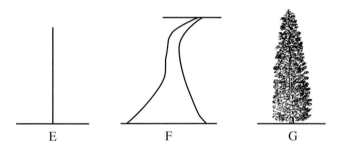

Figure 9.2 The horizontal–vertical illusion. The vertical line in E appears longer than the horizontal line. This might be because it is taken as representing a path receding into the distance on flat ground, as in F. It is then perceptually expanded to compensate for the foreshortened image. Experience of flat open environments might bias observers towards this interpretation. Experience of dense buildings or forests might instead bias the observer to interpret the figure as upright object, as in G, in which case no perceptual expansion is needed. (From Ross 1974, p. 60, with permission of the publisher.)

whether such differences are related to differences in depth perception and size constancy, or to educational or other reasons.[23]

Developmental studies

Another possible source of evidence for perceptual learning would be the development of the moon illusion in children. If the illusion were to increase or decrease with age, the trend would lend some support to the role of experience – though the data might not differentiate between horizontal and vertical experience. Experimental evidence would also be needed on the development of size constancy in both vertical and horizontal orientations.

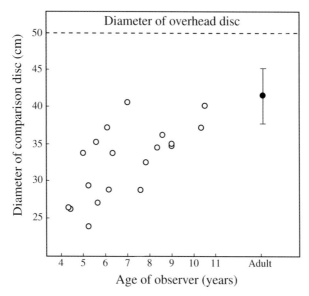

Figure 9.3 The effect of the observer's age on the matched size of an overhead disc of 50 cm diameter with a variable horizontal disc at the same distance. The matched size was too small, but grew with age, reaching the adult level by age 11. This implies a diminution of the moon illusion with age. (Redrawn from Leibowitz and Hartman 1959a.)

The first developmental study of the illusion appears to be that of Leibowitz and Hartman (1959a), and they found a *decrease* with age (Fig. 9.3). The experimental method unfortunately makes the results uninterpretable. The authors tested 19 children aged 4–11 years and 19 adults in an outdoor experiment, requiring them to match the 'apparent size' of an overhead disc suspended from a building at a distance of 26 m with that of a variable-sized horizontal disc at the same distance. They also conducted an indoor experiment in a darkened theatre, where they tested ten children and ten adults with spots of light viewed at a distance of 11 m (copying the method of Schur 1925, described in Chapter 11). The matched size of the horizontal disc was always smaller than that of the vertical, indicating horizontal enlargement. However, the difference was larger in young children than adults, and decreased to the adult level by 10–11 years of age. One might conclude that, if experience is a factor, its effect has already reached its maximum by about age five. However, Leibowitz and Hartman took the alternative view that the illusion was due to inexperience of vertical viewing, and that the reduction with age was caused by increased vertical experience and thus increased size enlargement in the vertical domain. The above argument is unsatisfactory, since the illusion consists mainly of horizon enlargement rather than vertical diminution, with respect to true angular size (Chapter 4). It is not surprising that the authors' conclusion drew several criticisms, and an exchange of views followed in the literature.[24]

About 30 years later Hershenson (1989a) and Gilinsky (1989) came up with a different type of explanation for the reduction with age: they proposed that it was related to an increase in body height rather than to experience. Da Silva (1989) suggested, perhaps more plausibly, that the trend was due to cognitive processes. Indeed, since the most obvious age trend occurred with the youngest children, one wonders whether they were fully able to understand the instructions and perform the experiment. Moreover, the 'error of the standard' is known to be important (Chapter 4): if the vertical and horizontal objects had been interchanged as standards, would the developmental trends have been the same?

One further developmental study of the moon illusion was reported by Thor, Winters, and Hoats (1969), and the results showed no age trends. They tested 128 children with a forced choice method, and found the illusion fully developed by age five. We describe this experiment in more detail in Chapter 11, along with other experiments on the angle of regard. No other developmental studies seem to have been conducted which might clarify the difference in size perception between horizontal and vertical viewing.

The evidence regarding the general development of size constancy in childhood is also unsatisfactory. There are many anecdotal reports of changes in size perception with age. For example, Helmholtz (1856–66/1962, p. 283) described his own experience as a boy:

> Incidentally, this relation between distance and size is something that can only be acquired by long experience, and so it is not surprising that children are not very proficient at it and are apt to make big mistakes. I can recall when I was a boy going past the garrison chapel in Potsdam, where some people were standing in the belfry. I mistook them for dolls and asked my mother to reach up and get them for me, which I thought she could do. The circumstances were impressed on my memory, because it was by this mistake that I learned to understand the law of foreshortening in perspective.

It is curious that Helmholtz should describe this incident of toy-like perception when looking upwards as an example of inexperience, when a few pages later (p. 291) he describes the moon illusion as due to perceptual learning about the probable distance of overhead objects. But such confusions are typical of this area of study.

Other anecdotal reports of childhood size perception focus on remembered size and its relation to body size. Many of us have the experience of returning to a childhood scene when an adult, and finding the buildings and people much smaller than we remembered.[25] This suggests that size scaling changes with body height, but it gives no insight into size scaling at different viewing distances.

Studies of infants show that size and depth perception develop over the first few months of life, along with the neurophysiological development of the visual system.[26] Experiments suggest that older infants have quite good linear size constancy at short viewing distances, but it is unclear whether this is a result of visual experience or of neurophysiological maturation. Older children can be tested with further viewing distances, using simple versions of adult test methods. Most such studies have involved only horizontal viewing, and they provide inconclusive evidence for a growth of size constancy.[27] One is left with the suspicion that any age trends for size constancy are just as fragile as those for the moon illusion.

It is interesting to compare the above studies with those for geometrical illusions of size. Not surprisingly, there is disagreement about the direction of any age trends for various illusions. However, many authors report that the Müller–Lyer arrowheads illusion decreases throughout childhood, while more conflicting trends are reported for the Ebbinghaus circles illusion (Fig.10.4) and the Ponzo converging lines illusion (Fig.10.5).[28] A decline with age makes it unlikely that perceptual learning is a factor, and more likely that cognitive factors are responsible.

The further-larger-nearer hypothesis

Many authors have attempted to explain why the moon appears nearer rather than further on the horizon. One type of solution is to say that it appears larger on the horizon for some reason other than its perceived distance, and that its enlarged perceived angular size causes it to appear closer through size–distance invariance. Such an argument departs from the idea discussed in the present chapter that the illusion is due to an error of distance judgement.

A different and paradoxical solution is to say that the horizon moon is seen as both further and nearer at the same time. We shall refer to this as the *further-larger-nearer* hypothesis. Reimann (1902a, 1902b) appears to have been the first to suggest this type of explanation, although it is not clear whether he did so intentionally. He stated that the flattened appearance of the sky causes the horizon moon to be seen as more distant and therefore (owing to size–distance invariance) as larger. He then argued that this greater perceived size causes the horizon moon to be consciously perceived as closer – 'floating' before the sky. The paradoxical nature of this argument was scathingly pointed out by Zoth (1902). However, his criticism had the unexpected effect of propagating the hypothesis rather than eliminating it. Claparède (1906a, p. 132) took up the idea and made it into an explicit theory (although he did not support it). He suggested that the distance theory could be saved if one assumed that two size–distance judgements of the moon are made. The primary one, made subconsciously, is that the horizon moon is further and therefore larger. This leads to the secondary inference that the enlarged moon must be closer. These inferences occur simultaneously, since the impressions of size and distance are immediate. Whether such contrary inferences could really be made at the same time by the same psyche seemed problematic to Claparède, even if they were made at different levels of consciousness.

This theory illustrates again the confusion resulting from the different meanings of the term 'perceived size' (Chapter 3). The first inference assumes that perceived size is analogous to linear size (which increases with distance at constant visual angle); and the second that it is analogous to angular size (which increases when an object draws nearer). The theory in this form is inadequate, because the operational meaning of 'perceived size' is changed dramatically during the argument.

The idea of unconscious inference appears to be based on John Locke's theory of unconscious perception. Locke thought that the ideas we derive from our sensations are often altered by our 'judgement', without our awareness. Although his theory was rejected during the eighteenth century by the French philosopher Etienne de Condillac, it survived. Since the late nineteenth century it has usually been attributed to Helmholtz.[29]

The size–distance paradox has often been found with changes in the vergence angle of the two eyes when stimuli of the same angular size are viewed in reduced cue situations: close vergence may make an object appear both smaller and further away, and far vergence has the opposite effect. Effects of this type have been reported by many authors.[30] Wheatstone (1852) was one of the first to investigate the phenomenon. He found consistent changes in perceived size, but very variable effects on perceived distance. He denied that convergence first determined perceived distance, and that perceived size was scaled according to this distance. Instead, he argued:

> From the experiments I have brought forward, it rather appears to me that what the sensation [of convergence] immediately suggests is a correction of the retinal magnitude to make it agree with the real magnitude of the object, and that distance, instead of being a simple perception, is a judgement arising from a comparison of the retinal and perceived magnitudes. However this may be, unless other signs accompany this sensation the notion of distance we thence derive is uncertain and obscure, whereas the perception of the change of magnitude it occasions is obvious and unmistakeable.

An equally confusing explanation was put forward by Myers (1911) in relation to size–distance effects in a fog (Chapter 6) and to geometrical illusions. It had been argued that some geometrical patterns contain depth cues, which alter the perceived depth and thus the perceived (linear) size of certain parts of the figures.[31] It was soon pointed out that conventional size–distance invariance breaks down in these illusions, since the perceptually enlarged parts appear to be either at the same distance or nearer than the perceptually diminished parts. To account for this, Myers (1911, pp. 282–3) wrote:

> In the latter case [a foggy atmosphere] … and in the suggestions of perspective in a drawing – it is the apparent size which determines the apparent distance. Yet primarily, the apparent size must be dependent on some unconscious influence of distance. Possibly we have here a schema. … or unconscious disposition, in regard to the distance of objects; and when this schema undergoes change, it manifests itself in consciousness by effecting a change in apparent size, whereupon the apparent size determines our awareness of the distance of the object.

Many recent authors have also put forward versions of the further-larger-nearer hypothesis. Dees (1966b) proposed that the horizon moon appears further away through monocular depth cues such as linear perspective and interposition; size–distance scaling makes it appear larger; then the larger size makes it appear closer. A more sophisticated argument was put forward by Kaufman and Rock (1962a, 1962b) and Rock and Kaufman (1962), who distinguished between two types of distance perception – the *registered* distance (which may be unconscious and is used to determine perceived size), and the *judged* or apparent distance (which reaches consciousness after the perceived size has been taken into account). In their 1989 account, Kaufman and Rock maintained that the registered distance of the horizon moon was consistent with the consciously perceived distance of the horizon, and that the judgement of the moon's distance was secondary and was distorted by the moon's perceived size. These authors stated that many of their subjects spontaneously reported that the horizon moon was 'closer because it was larger'. Kaufman recently conducted a new experiment involving binocular disparities (Chapters 7 and 11), together with his son (Kaufman and Kaufman 2000). Five subjects, including the elder Kaufman, viewed two stereoscopic pairs of artificial moons projected over a natural landscape: they adjusted the lateral separation of one pair so that the nearer fused moon appeared to be at half the distance of the further fused moon on the horizon. They repeated the measures with the apparatus raised at an angle of 45 degrees. The relative disparities were significantly smaller for the horizontal than the elevated settings, and the authors argued that this implied that the perceived distance was greater for horizontal viewing. They concluded that 'the horizon moon is seen as larger because the perceptual system treats it as though it is much further away'. However, they went on to hedge their bets about the causal role of perceived distance by saying 'Thus the term *apparent* in so-called apparent-distance theories is inappropriate. Rather, we suggest that the physical cues to distance affect both perceived distance and perceived size.'

Several other authors have attempted to distinguish between automatic perceptions and cognitive judgements. Gogel (1974) and Gogel and Mertz (1989) argued that the 'equidistance tendency' mainly determines the perceived distance of the horizon moon, so that it is seen to be close to the horizon; but the 'egocentric reference distance' mainly determines that of the zenith moon, so that it is seen to be very close. The 'reported distance' of the horizon moon is then cognitively determined by the judgement that the moon is a large 'off-sized' object.

The idea that the size–distance paradox is cognitive rather than perceptual was also supported by the experimental evidence of Mon-Williams and Tresilian (1999). They altered vergence by placing a prism in front of one eye. Their findings were similar to Wheatstone's: when the observers converged to a closer distance than the target, they reported that the target appeared smaller and further; and the opposite effects held for far vergence. However, the distance effects only occurred in verbal reports: when the observers reached with their finger (hidden from view) to the perceived location of the target, they reached closer for near vergence and further for far vergence. The authors interpret this to mean that the motor system receives accurate but 'cognitively impenetrable' visual information about distance, while the cognitive visual system relies on various cues including perceived size.

There are difficulties in accepting that the horizon moon can be 'perceived' as both further and nearer at the same time, even at different levels of processing.[32] The concept of registered distance has been given empirical support by some non-verbal measurements (e.g. Roscoe 1989; Mon-Williams and Tresilian 1999; Kaufman and Kaufman 2000). However, other experiments involving the manual adjustment of the distance of targets (described in Chapter 12) confirm the verbal reports that horizontally viewed targets appear both nearer and larger than vertically viewed targets. The topic remains controversial, and we discuss it further in Chapter 12 in relation to the idea that there are two different 'visual streams' for perception and action. We turn now to some other theories involving more than one type of perceived distance.

Reference level theories

As described in the previous chapter, von Sterneck (1907, 1908) developed a quantified reference level theory of perceived distance, which he applied to the shape of the sky and other distances. A key point in the theory is that there exists a maximum perceivable distance, which varies with the illumination and other viewing conditions. A similar theory was developed by Gilinsky (1971, 1980), who named this distance the adaptation level (A), following Helson's adaptation level theory (1964). Her formula for the relation between perceived distance (d) and physical distance (D) was:

$d/D = A/(A + D)$.

This formula is identical to that of von Sterneck, when the relevant symbols are interchanged. Gilinsky claimed from her interpretation of various experiments that A varied between 0.5 and 91.5 m – a much shorter distance than von Sterneck's 200 m to 20 km. She argued that our visual space expands or contracts in different settings, distance adaptation varying with the presence of distance cues, the direction of regard, and the total distance range. These factors could account for the apparently flattened dome of the sky.

Gilinsky differed from many flat-sky theorists in stating that the perceived distance of the moon was not identical to that of the sky. Instead, she argued that the enlarged distance scale on the horizon 'moves past' the horizontal moon, while the contracted vertical scale moves in front of the zenith moon. She reanalysed the data of the two subjects in Holway and Boring's (1940a) experiment on the angle of regard, incorporating the calculated maximum perceived distance. She concluded that the perceived distance of the moon and that of the sky coincided at an angle of elevation of about 20 degrees, and that what she called the normal 'memory' size of the moon occurred at this viewing height.

The concept of remembered or recalled size was important to Gilinsky. In explaining the moon illusion (1980, p. 281) she wrote:

> What produces the change in perceived size? The answer must be that an observed size at an observed distance is compared with a recalled size – an internal referent established by the memory moon at normal viewing distance. This standard bends the perception back to the familiar 'true' size (S) – the equivalent, for the moon, of size constancy. The result is the dramatic variation in the perceived size in the moon. The horizon moon appears larger because A has a higher value; the zenith moon appears smaller because A has a smaller value.

She expressed the relation between perceived and true sizes and distances by the following formula:

$s/S = (A + \delta)/(A + D)$.

where s is perceived size, S is true size, D is true distance, δ is normal viewing distance, and A is the adaptation level (maximum perceivable distance). This formula is hard to interpret. The use of size ratios evades the question of whether sizes are intended to be angular or linear, and the idea of a 'normal viewing distance' is difficult to pin down. Gilinsky published her general theory of space perception in 1951, but the derivation and meaning of her equations were immediately criticized by Fry (1952) and Smith (1952). Her 1989 account remained basically unchanged.

A different reference theory was developed by Baird (1982). He maintained that the sky was seen as hemispherical rather than flattened. However, his main assumption was that 'the apparent size of the moon is compared against objects on the ground in the vicinity of the point at which a plumb line from the moon would reach the ground, with apparent size and distance linked by the size–distance invariance hypothesis' (p. 305). Thus the perceived size of the moon is related to the perceived size and distance of objects on the ground below it. Baird added that as the moon rose in the sky the influence of ground referents became less, and that of the sky more important: 'A major referent for the moon is the empty expanse of the sky … A perceptual combination of ground and sky referents is the cause of the moon illusion' (p. 305).

Baird went on to explain differences in the perceived distance of the moon by changes in its perceived angular size, the apparently enlarged horizon moon appearing nearer. Baird denied that his theory involved a 'judged' and 'registered' distance, maintaining that the perceived distance to the moon and the perceived distance along the ground involved two different *physical* distances. However, it is hard to distinguish this reasoning from the 'further-larger-nearer' theories.

Inverse transformation models

The simple form of the size–distance invariance hypothesis (as discussed in Chapter 3) can only apply to upright objects viewed at right angles to the observer. The usual formula does not deal with the size of objects lying on the ground or at a slant. Some more complex formula is necessary, and some of the possibilities are discussed by Schwartz (1994). A diagram giving some of the possible variables is shown in Fig. 9.4.

To determine the size (s) by triangulation, it is necessary to know not just the subtended visual angle (α) but also the two distances d_1 and d_2, or one of these distances and one of the other interior angles (β and γ).

All the possible formulae – of which classical size–distance invariance is a simplified example – imply that spatial perception consists in projecting the visual image out into a three-dimensional space in a geometrical manner (sometimes called an 'inverse transformation'). However, to achieve the correct projection, some aspects of the space must be determined by information other than the static visual image.

Inverse transformation models have been applied to size constancy and to the moon illusion. Baird later rejected his 1982 reference model in favour of a more general transformation model (Wagner, Baird, and Fuld 1989; Baird, Wagner, and Fuld 1990).

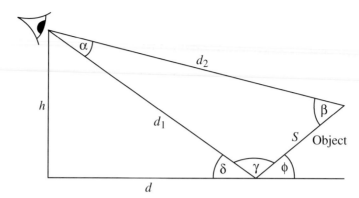

Figure 9.4 Sources of information for a general solution to the computation of size (*s*) by triangulation. The subtended angle (α) is assumed to be known. Other sources are the line-of-sight distances d_1
and d_2 and the interior angles β and γ; or the observer's eye-height (*h*), the ground distance to the base
of the target, and the target orientation (ϕ).

These authors give a diagram essentially similar to Fig. 9.4, though they differ from
Schwartz in assuming that different dimensions are known. In this case the geometry is
calculated from the angle subtended at the observer's eye (α), the observer's eye-height
(*h*), the distance along the ground to the bottom of the target (*d*), and the angle of
orientation of the target with respect to the ground (ϕ). This appears to be reducible to
Schwartz's case of knowing one distance and one angle in addition to the subtended
angle: indeed, the authors state that the major parameters of their model are perceived
distance and applied orientation. The relevance of this formula to the moon illusion is
that the applied orientation (ϕ) is smaller when the moon is near the horizon than near
the zenith, and this smaller angle enlarges the perceived (linear) size of the moon.
The applied orientation is smaller at the horizon because of the influence of the
flat ground. The authors do not claim that the perceived orientation of the moon
changes, but rather that there is a change in the inverse transformation used to
calculate size.

With this model, the authors do not attempt to explain the size–distance paradox
of the moon illusion. The perceived distance to the moon is said to be the same at all
orientations, and it is the applied orientation that changes. However, the trigonometry
of angles and lengths is inextricably mixed, and a change in applied orientation implies
a change of distance at one end of the target at least. Like all projectionist models, this
has to be classed as one of perceived distance.

We conclude that none of the perceived distance theories is able to account satisfacto-
rily for the size–distance paradox, at least within the context of classical size–distance
invariance. In the next chapter we turn instead to theories that involve only size
relationships.

Summary

The perceived distance of the celestial bodies is usually reported to be closer near the horizon than higher in the sky, and not further as required by classical size–distance invariance. The presence of intervening objects near the horizon has often been used to explain both the flattened dome of the sky and the supposed increase in the perceived distance of the horizon moon. Early 'tube observations' that cut out surrounding objects produced conflicting results, probably because of inadequate experimental methods. Modern experiments show that the presence of intervening objects, or terrain, can increase perceived size by about 34 per cent, but the amount varies with the effectiveness of the display. However, intervening objects may affect perceived size independently of perceived distance. Other perceived distance explanations involve the role of experience, either through the expansion of horizontal distances more than vertical distances, or through the contraction of vertical distances. There is no experimental evidence to distinguish between these possibilities, and any developmental trends are controversial. Some authors attempt to maintain the classical size–distance invariance explanation by the 'further-larger-nearer' hypothesis, in which two types of distance perception are specified (registered and estimated) and linear and angular size are confounded. This is unsatisfactory, and perceived distance cannot be used as the main explanation of the celestial illusion.

CHAPTER 10

Sizing up the moon

The most common explanation of the moon illusion given by lay people is the erroneous one of refraction. The next most common is that of relative size. For example, Ruggles (1999, p. 191), an archaeologist and astrophysicist, writes:

> Most of us share the perception that the moon appears particularly spectacular when near to (say within one or two degrees of) the horizon, and especially when it is close to small and distinct horizon features, such as distant mountains, buildings or trees. This is doubt-less due in part to the well-known effect that the proximity of the moon to terrestrial features makes it seem larger to the eye.

Many authors have attempted to explain the moon illusion as some type of relative size, and they usually mention the importance of objects on the horizon. To illustrate the idea, we show a card with the full moon above a silhouette of chimney-pots (Fig. 10.1).

The explanations take various forms, which tend to merge into one another. The simplest account is that of *size contrast*, in which the difference in real angular size between adjacent objects is perceptually enhanced: thus an object of a given angular size

Figure 10.1 A Christmas card by Rosemary Smith showing the full moon above a silhouette of chimney-pots. (Reproduced by permission of the artist.)

will appear large when close to objects of small angular size, and small when close to objects of large angular size. The low moon appears large in relation to other objects of smaller angular size on the horizon. This account does not involve any form of size–distance invariance.

A more complicated account is that of *size assimilation*, in which an object near the horizon is seen in comparison with large objects of known linear size. Assimilation theories involve two stages: first, terrestrial objects near the horizon are perceptually enlarged by some size scaling process; second, celestial objects near the horizon are assimilated to the same scale as terrestrial horizon objects. These theories may rely on size–distance invariance for the enlargement of terrestrial objects, or they may rely on other size cues over the terrain.

It should be noted that while contrast and assimilation are apparently opposite explanations, they achieve the same end result. Unfortunately, few authors have described their theories with any degree of precision, and it is often hard to classify them. Indeed, some theories seem to be combinations of contrast and assimilation. We shall first discuss contrast theories, then assimilation theories, and finally other miscellaneous size theories.

Size contrast

Angular size contrast

It is not clear when the idea of angular size contrast (unrelated to objects of known size) first arose. Unlike the idea of size–distance invariance, it is not supported by an unbroken mainstream literary tradition: instead, it slips into the mainstream from various different sidestreams. As we noted in Chapter 7, Leonardo da Vinci (1452–1519) commented on the role of relative size in countermanding other possible size effects,[1] but did not give a detailed explanation. 'Relative size' or 'size contrast' form the basis of many geometrical illusions, and these concepts were made clearer when geometrical illusions were studied seriously in the nineteenth century. Hering (1834–1914) – the nativist competitor to Helmholtz's empiricism – held such a theory;[2] and the idea was later taken up by Gibson (1950, 1966) for his account of size and distance perception.[3] Rock and Ebenholtz (1959) approached the matter from a different theoretical background: they were influenced by experiments on brightness constancy, which showed that perceived brightness depends on the ratio of the luminance of a surface to that of its surround.[4] They applied a similar idea to the effect of frames on judged line length. Their observers viewed two luminous and widely separated rectangles in the dark, each containing a vertical line, and they adjusted the length of the line in the larger rectangle to match that in the smaller rectangle. The matched length of the line diminished as its surrounding frame was increased in size, though the effect was less than that required for strict proportionality. The framing effect was greatest (73 per cent of full proportionality) for relatively small frames, but reduced to 43 per cent when the frame was eight times the height of the line. Similar, but less systematic, measures were made previously by Cornish (1935, p. 48). He drew a house which he viewed through a window frame, and he varied his distance from the window. The angular size of the window was 14, 33, or 100 degrees, and the corresponding size of the drawing of the house was 5.8, 4.2, and 2.5 inches. Thus when the angular size of the frame was enlarged by a factor of 2.4, the depicted size of the house was reduced by a factor of 1.38 (about 58 per cent of full proportionality); and when the frame was enlarged by a factor of 7.1, the depicted size of the house was reduced by a factor of 2.32 (about 31 per cent of full proportionality).

A photograph of the setting sun framed between houses and trees is shown in Fig. 10.2. In this scene the sun appeared to three observers to be only slightly enlarged – unlike the view of the sun across the open sea (Fig. 10.3) which appeared very large. Framing does not therefore seem to contribute much to the celestial illusion.

A typical size contrast illusion is shown in Fig. 10.4. The figure is known as the Titchener or Ebbinghaus circle illusion, Titchener having credited it to Ebbinghaus (Titchener 1901). The size effect in this figure is quite small, perhaps 10 per cent; but some other geometrical illusions of size have been reported to give effects as large as 20–34 per cent.[5] Such effects could therefore contribute significantly to the moon illusion. Geometrical illusions are similar to the moon illusion, in that neither of them obey classical size–distance invariance: segments of the same angular size appear to differ in size while remaining in the plane of the paper.[6]

It is often argued that size contrast does indeed contribute to the moon illusion, because the moon is viewed in relation to objects of small angular size on the horizon, but in isolation (or perhaps in relation to the large framework of the sky) when in the zenith. This idea formed part of the explanations proposed by Logan (1736), Huxley

Figure 10.2 The setting sun viewed between trees and houses across Greenwich Park, London – near the Royal Observatory and the meridian line. The sun did not appear much enlarged. (Photograph by Helen Ross.)

Figure 10.3 The rising sun viewed across the harbour at Dover. The sun appeared very large, despite its small size in the photograph. (Photograph by Helen Ross.)

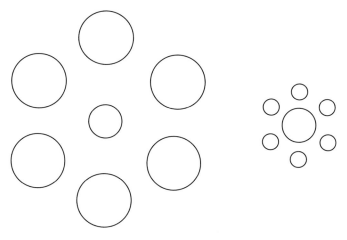

Figure 10.4 A size contrast illusion, known as the Titchener or Ebbinghaus circles. The centre circle on the right, surrounded by small circles, appears larger than the one on the left surrounded by large circles.

(1885), and some twentieth-century authors.[7] Against this it can be said that the raised moon is often surrounded by small stars, which should act to increase its perceived size. Another potential negative point is the 'tree test' – the claim that the moon (at any elevation) is not significantly enlarged when viewed through the foliage of a tree.[8] Like other 'critical' observations, this one is debatable. For example, the horizon moon viewed through the branches of nearby trees may appear small when apparently framed at the same distance as those trees (as described above for the sun in Fig. 10.2), but large when apparently located behind further terrain visible through the branches.

Various authors have conducted experiments on the effects of the terrain on the artificial moon illusion. These experiments were discussed in Chapter 9, since most authors argued for an explanation in terms of perceived distance. However, the experiments could equally well be classified as examples of the role of contrast or assimilation. The largest effect was reported by Rock and Kaufman (1962): they found that, when two low artificial moons were compared, the moon with the terrain visible was judged on average 34 per cent larger than the moon with the terrain obscured – a value similar to that obtained for the comparison of their horizon and zenith moons.

If size contrast is an important factor, its effects should be visible in suitable drawings of the moon and terrain, just as it is in geometrical illusions. Coren and Aks (1990) measured the moon illusion in pictorial arrays, and found that horizon moons were indeed judged larger than elevated moons of the same size, but only by about 2 per cent on average. They then used various different displays and measurement techniques, and found mean illusions ranging between 11 and 36 per cent for optimal displays. The average effect for pictorial arrays is thus smaller than for the real terrain viewed through a moon machine, or for other measures of the natural moon illusion. Coren and Aks argued that pictorial elements could account for between 22 and 72 per cent of the full natural illusion. They also argued that the effect depended on the strength of the depth cues in the displays. It is interesting to note that inverting the display reduces the illusion to slightly less than half its size, both with the pictorial illusion (Coren 1992), and with moon machines and real terrain (Rock and Kaufman 1962). This suggests that size contrast is not the only factor, but that height in the visual field, or some higher cognitive processes, are also important. Height in the visual field can be a cue to depth, which may explain why inverted pictures convey a reduced impression of depth. The above-mentioned authors used the reduced impression of depth to explain why inverted scenes reduce the moon illusion. Alternatively, it could be argued that height in the visual scene directly affects size perception, independently of any distance effects.[9]

It should be noted that angular size contrast is a local process that only works for images that are close to each other in the visual field. To contribute to the moon illusion, other objects on the horizon would have to subtend an angle of less than half a degree. Yet the illusion occurs when the horizon is empty, or when it contains objects of larger angular size. Larger objects in the foreground should act to reduce the illusion, if their angular size were to be compared directly with that of the moon. Since the foreground terrain does contribute to the illusion, it must do so by its contribution to the gradient of scaling over the whole scene. The role of local size contrast thus seems to be minimal.

The expanse of the sky

A variation on the theme of relative size relates the illusion to the expanse of sky within which the moon is viewed. The horizon moon looks large compared to the small extent

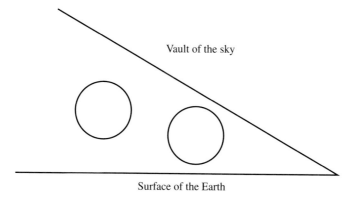

Figure 10.5 Ponzo's 'angular contrast' illusion as an explanation for the celestial illusion. The circle on the right appears larger than that on the left, because it occupies a greater proportion of the space between the enclosing lines. The circle on the right could represent the moon on the horizon, enclosed between the line of the sky and the line of the earth's surface. (After Ponzo 1913.)

of sky between it and the earth's horizon, while the elevated moon looks small compared to a wide expanse of sky. This suggestion seems first to have been made by Lühr (1898). It was criticized along with the other size contrast theories, the 'tree test' (described above) being used against it.

Ponzo (1913) suggested that the moon illusion was a special case of what he called the 'angular contrast' illusion (now known as the Ponzo illusion). His illustration is given in Fig. 10.5. He proposed that, on the horizon, the celestial bodies were contained within the angle formed by the junction of the earth's surface with the falling arc of the sky, while in the zenith they were free of this framework. Variations in the strength of the illusion could be caused by clouds and vapours, which change the perceived angle that the celestial vault makes with the ground. An analogy with the Ponzo illusion was recently proposed by Wenning (1985), though he ascribed both illusions to perceived distance rather than to size contrast.

Patrick Rizzo, an amateur astronomer, reinvented the explanation based on the expanse of the sky, and described it in various bulletins in the 1960s; consequently Restle (1970b) credited him with the idea.

Adaptation level theory

Some authors have couched the various relative size explanations in the mathematical terms of adaptation level theory (after Helson 1964). One of these was Restle (1970a), whose theory is one of size alone, with no hint of taking distance into account. It differs from the adaptation level theories of Gilinsky (1971, 1980) and Baird (1982), which are grounded in distance theories (Chapter 9).

Restle (1970a) held a sky expanse theory similar to that of Lühr (1898), but produced a mathematical account. Restle argued that size was judged relative to the adaptation level (A) for visual extent, which was derived from the weighted geometric mean of all relevant objects or extents. The moon appears small near the zenith because A is large: A is large because the angular separation between the horizon and the moon (up to 90 degrees) is an extent entering the equation. The moon appears large near the horizon because A is small: A is small even when the horizon is featureless, in which case it is determined by the very small angular gap between the moon and the horizon. The size of this angular separation is thus an important factor in the equation. Restle explained the reduction in the moon illusion in Kaufman and Rock's inverted display by claiming that an inverted scene makes the observer take in a wider variety of objects, thus increasing the adaptation level.

Restle went on to make quantitative predictions, based on results from his laboratory on the effects of surrounding frames on perceived size. However, these experiments used small separations of the stimulus and the adjacent frame, and it is not clear that the results can be extrapolated to very large separations, such as that between the zenith moon and the horizon.

Divided space

Intervening objects may expand perceived distance (Chapter 9). Here we consider their possible role in expanding perceived size. Goüye (1700) suggested that terrestrial objects seen against the moon's disc on the horizon make the moon seem larger because they divide it into parts, and a divided space always looks larger than the same space when undivided. On this basis, just as on relative size accounts, the moon should appear larger when viewed through trees. There is little evidence that it does so.

The phenomenon underlying this explanation later came to be known as the Oppel–Kundt illusion, or the Botti illusion.[10] The illusion figure (Fig. 10.6) produces a maximum size enlargement of about 8 per cent when there are nine bars, the effect being independent of the retinal size of the separation between the bars.[11]

Divided space cannot be a complete explanation of the moon illusion because the effect is much too small, and because the illusion exists when the moon is clear of horizon objects. However, it could well be a contributory factor when the moon is low on the horizon, or when wisps of cloud cross its circumference.

The horizon line

The line of the horizon is of particular significance for orientation to the environment, and for judging the slopes of hills and roads.[12] More significantly for the moon illusion,

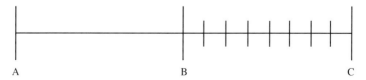

Figure 10.6 The Oppel–Kundt illusion. The separation between B and C appears greater than between A and B. (After Oppel 1855).

it has been claimed that the *horizon ratio* provides important information for judging the size of terrestrial objects. This theory is based on the laws of perspective, and was originally stated or implied by Euclid in his *Optics* around 300 BC. It was restated by Gibson in 1950, and further developed by Sedgwick (1973, 1986). The horizon ratio principle states that if an observer is standing on flat terrain, a point on an object that intersects the horizon will be one eye-height above the ground. Terrestrial objects of the same height are all cut by the horizon in the same ratio, regardless of distance, as is the case for the trees in Fig. 10.7.

The theory has the attraction of allowing all objects that appear to be divided by the horizon line to be scaled for size in terms of the observer's own eye-height. Other objects that do not touch the horizon could then be scaled in relation to those that do. A horizon ratio account of the moon illusion was briefly described by Warren and Owen (1980), but was not fully developed.

One difficulty with the horizon ratio concept is that the ratio could only provide veridical size information when the horizon is very distant and the observer is on flat ground. For much terrain, a line of sight to a terrestrial horizon does not lie parallel to the ground. If the effective horizon is raised (as when looking towards gently rising hills), underestimation of size should occur; and when it is lowered (as when looking to the ground below buildings), overestimation should occur. However, the evidence suggests that underestimation occurs when looking both up and down (Chapters 11 and 12). Moreover, mountainous terrain can cause misjudgements of the elevation of the horizontal.[13] The current status of the horizon ratio principle is also equivocal.[14]

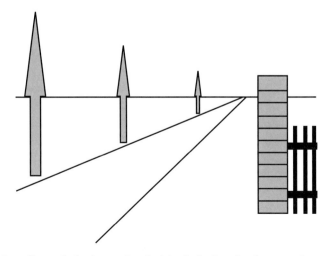

Figure 10.7 According to the horizon ratio principle, the horizon line intersects the trees and the pillar at a height equal to the observer's eye level. Objects of the same linear size have the same horizon ratio. Since a smaller proportion of the pillar is above the horizon, the observer perceives it as smaller than the trees.

Size assimilation

Comparison with objects of known size

The idea that some reference scale is needed to determine perceived size is not new; but, like the idea of angular size contrast, its origins are obscure. We quoted the rather tentative views of Plotinus (*c.* 203–70) in Chapter 3. However, Benedetto Castelli (1578–1643) seems to have been the first to ascribe the celestial illusion to comparison with objects of known size. He wrote about it in the *Discorso sulla vista*, composed in 1639 and published posthumously in 1669. In Section XXIV he noted that the Big Dipper appeared larger on the horizon, even though it subtended the same angular size as in the zenith:

> After having reflected much and carefully it came to my mind that this business of the large and the small is by our mind managed always in relation to some other magnitude better known to us than that which is the magnitude of the object about which we must form an idea of whether it is large or small. In our case, as we look at the regions of the Heaven that are located near the zenith, we are led to compare them and to relate them to the summits of the roofs of our buildings, for we have nothing else near the zenith with which we can compare them. For this reason the constellation of the Dipper, when it is looked at in similar circumstances, will seem to us as occupying a portion of space equivalent to that of the top of a house or of a temple, which space is well known to us and does not attain more than a few tens of yards. In this manner we can form a rough idea of the size of the Dipper which we judge and estimate as [rather] small. But when we see it near the horizon we compare it with the long chains of mountains and with the wide countryside, which we know very well are sometimes tens of miles. In such a case we judge the Dipper to be much larger than what we had thought in the other disposition. In this way the object whose magnitude must be judged is appraised by us sometimes larger and sometimes smaller as we compare it with different magnitudes.[15]

Castelli was perhaps the first to experiment on the effects of the terrain and relative size. He went on to describe how a party of people were travelling along the Tiber in a carriage, and saw the full moon rising over the Aventine Hill on the other side of the river. When the brim of his hat covered the view of the hill, but the moon could be seen over the hat brim, the moon appeared very much smaller than when seen over the hill. This was observed by all members of the party. Castelli thought the effect was due to the small size of the hat brim in relation to the hill.

Castelli appears to have held mixed views, because he also believed that the perceived distance of the horizon was critical: 'we reckon those parts which are near the vertex as very near to the eye and those which are placed along the horizon as very far away'.[16] Castelli's change of view came about as a result of discussions with his teacher, Galileo, who convinced him of the importance of the size–distance invariance explanation.

Molyneux (1687) wrote a review of the moon illusion, criticizing all known theories. He condemned the idea of comparison with known objects, which he ascribed to Descartes:

> for the Moon (says he) being nigh the *Horizon*, we have a better opportunity and advantage of making an estimate of her, by comparing her with the various objects that incur the

sight, in its way towards her; so that tho we Imagine she looks bigger yet tis a meer Deceipt: for we only think so, because she seems nigher the tops of Trees or Chimnys or Houses or a space of Ground, to which we can compare her, and Estimate her thereby; but when we bring her to the Test of an Instrument that cannot be deluded or Imposed upon by these appearances, then we find our Estimate wrong, and our Senses deceived. These thoughts, my-thinks, are much below the Accustomed Accuracy of the Noble Des-Cartes; for certainly if it be so, I may at any time increase the apparent Bigness of the Moon, tho in the Meridian; for it would be only by getting behind a Cluster of Chimnys, a Ridg of a Hill, or the top of Houses, and comparing her to them in that posture, as well as in the *Horizon*: besides if the Moon be look'd at just as shee is Rising from an *Horizon* determined by a smooth Sea, and which has no more Variety of Objects to compare her to, than the Pure Air; yet she will seem bigg, as if lookt at over the Rugged top of an uneven town or Rockey Country. Moreover, all Variety of adjoyning objects may be taken off, by looking through an empty Tube, and yet the deluded imagination is not at all helped thereby (pp. 316–17).

Molyneux appears to be misquoting Descartes. In the Sixth Discourse of *La dioptrique* (1637a) Descartes mentions only the effect of intervening objects upon perceived distance. However, he was a voluminous writer, and may have mentioned relative size elsewhere. Alternatively, Molyneux may have come across the explanation in some other source, and failed to realize that it differed from that of Descartes. He clearly did not regard it as a new idea. Molyneux himself was unable to reach a conclusion concerning the cause of the illusion.

Logan (1736) reasserted the importance of comparison with objects of known size. He stated that the horizon sun appeared even more enlarged than usual if viewed between tall trees:

Now the Reason of this is obvious, *viz.* that being well acquainted with Trees, the Ideas of the Space they take up are in a Manner fix'd, and as one of those Trees subtends an Angle at the Eye, perhaps not exceeding two or three Seconds, and would scarce be distinguishable, were it not for the strong Light behind them, the Sun's Diameter of about thirty Minutes takes in several of them, and therefore will naturally be judge vastly larger. Hence 'tis evident, that those bodies appear greater or less, according to the objects interposed or taken in by the Eye on viewing them. And to this only is that Phenomenon to be imputed.

I am sensible this Method of arguing is not new, yet the observations here given may probably tend to illustrate the Case beyond what had been advanced on the Subject (p. 405).

Trees on the horizon were commonly given as a reason for the sun's enlargement. Le Conte (1881, p. 159) stated that the sun or moon's estimated size usually varied from that of a saucer to the head of a barrel (see Chapter 3):

But under peculiar conditions we imagine them much larger. For example, a pine-tree stands on the western horizon about a mile distant. I am accustomed to judge of the *size* and distance of trees. This one seems to me at least 20 feet across the branches. The evening sun slowly descends and sets behind the tree. *It fills and much more than fills its branches.*

Comparison with terrestrial objects of known size, or of perceptually enlarged size, was also given as the main explanation by many later authors.[17] Filehne (1894) and

Müller (1906, p. 309) criticized the size comparison explanation, claiming that it failed for the constellations. However, this objection might not be fatal, since the constellation illusion is usually smaller than the moon or sun illusion. Filehne also claimed that size comparison did not explain the moon illusion over the sea. The latter example is often regarded as a test case for the 'known size' account, since an empty sea contains no objects of known size (Claparède 1906a). Solhkah and Orbach (1969) argued that an extent of water may be regarded as an intervening familiar object: if the surface texture is fine, the horizon will be judged distant and the moon large; while if the texture is coarse, as when the observer's head is low in the water, the horizon will be judged nearer and the moon less enlarged. This is, however, a perceived distance explanation rather than a relative size explanation. Just as with the 'tube' experiments, there is disagreement as to whether the moon illusion is indeed reduced over the sea. For example, Molyneux (1687) claimed that it was not reduced, while Mayr (1904) and Minnaert (1954) stated that it was slightly reduced. Many people report seeing a large illusion over a totally empty sea. Often, however, some adjoining land or other objects are visible, giving scale to the scene. For example, Fig. 10.3 shows a photograph of the rising sun viewed over the sea across a harbour, and taken with a standard lens. The sun appeared very large to the observer (HER), and seemed to be quite isolated; but the photograph shows that other objects were present, and that the sun subtended about the same visual angle as a distant ship. The photograph also reveals a lot of texture from changes in luminance.

Size scaling in terrestrial and reduction viewing

The authors mentioned above all had something to say about the role of horizon terrestrial objects in the illusion, but with many variations and interactions. Another variation on assimilation theory proposes two different mechanisms of size perception, one operating under terrestrial conditions and the other under reduction conditions.

Early exponents of the latter view were Mayr (1904) and Claparède (1906a). Mayr argued that distant objects are usually seen as quite large, despite their small angular size. The horizon moon is perceived as a distant object in a terrestrial setting, and is therefore seen as large. The elevated moon, on the other hand, is seen in unusual circumstances and at an indeterminate distance, so the normal perceptual enlargement does not occur. Aerial perspective makes the horizon moon look more like a terrestrial object, and enhances the enlargement. The illusion is usually more pronounced for the moon than the sun, because the dim moon looks more like a terrestrial object: the sun illusion is pronounced only when the sun appears dim and reddened. The moon illusion occurs when no terrestrial objects are visible (for example over the sea), because the moon appears to be a distant terrestrial object. However, the illusion is most pronounced when comparison objects are also visible.

The idea that size perception is different under reduction conditions was expressed more explicitly by Grijns (1906). He noted that a candle flame seen through a pinhole appears smaller than usual, and presented this as similar to the moon illusion. Haenel (1909, p. 189) took the argument a little further. The perceived size of the horizon moon, like that of terrestrial objects seen in three-dimensional space, is analogous to linear size. However, the perceived size of the elevated moon is determined by its angular size alone. The factors determining the moon's perceived size are therefore not the same in the two conditions.

A somewhat similar view was put forward by Gregory (1970, pp. 104–5;1998, pp. 230–1). He suggested that the elevated moon takes on an average (or *null*, or *default*) assumed size and distance for its retinal image size; whereas the horizon moon is automatically scaled by the nearby ground cues such as perspective. Because the horizon moon is seen against an untextured sky background, its perceived distance is not attached to that of the sky, and its enlarged size makes it appear near. The horizon moon thus looks both nearer and larger than the elevated moon, in contradiction to classical size–distance invariance; it is therefore an example of *primary* or *bottom-up* scaling, like the geometrical illusions, in which size is scaled independently of perceived distance.

Reduction theories remain problematic. Why does a cloudy and textured sky fail to turn the raised moon into a 'terrestrial' object? What determines the perceived size of a reduction object, and is it always the same? Angular size matching experiments suggest that the perceived size of the elevated moon is variable: some authors claim that it appears equal to nearby objects of the same angular size, and others, such as Holway and Boring (1940a), that it appears much larger (Chapter 4). It is unclear, however, whether the variability reflects genuine differences in perception, or merely differences in measurement procedures. There also seems to be no evidence for any change in the nature of size perception between the horizon moon and the elevated moon. Indeed, the evidence from verbal reports suggests that – within the framework of size–distance invariance – there is no change: people who are prepared to estimate the size of the horizon moon in linear units (as discussed in Chapter 3) do the same for the elevated moon. Given the absence of evidence for a discontinuity, it is better to adopt theories that account for size scaling over the whole of the visual scene in a consistent manner.

Other size theories

Kinetic and static size scaling

A very different two-process theory was suggested by Hershenson (1982, 1989a). He noted that the usual size–distance relationship seemed to be reversed both in the moon illusion and in the case of the rotating spiral. A spiral rotating in one direction appears to expand and approach, while rotating in the other direction it appears to contract and recede; after viewing one direction of rotation for about 20 seconds, a stationary spiral appears to change in the opposite manner (Szily 1905). Hershenson referred to this larger-nearer relationship as *kinetic* size–distance invariance, which he contrasted with the further-larger relationship or *static* size–distance invariance (which we have called *classical* size–distance invariance). He suggested that there was a 'loom-zoom' system which, when stimulated by a symmetrical expansion or contraction pattern, produces the perception of a rigid object moving in depth. He further suggested that a moon–terrain configuration stimulates the loom-zoom system, causing the apparent increased size and near distance of the horizon moon. However, it is not clear why a *static* moon–terrain configuration should stimulate a *kinetic* system, or why kinetic size–distance effects should be considered to be different in kind from static effects. There need be no conflict between the two systems if the kinetic system is understood to refer to angular size and the static system to linear size.

A failure of size constancy

Many authors regard the moon's enlarged appearance on the horizon as an example of normal size constancy. In this case its small appearance in the zenith is in need of explanation, and represents some breakdown in size scaling. Since its small appearance is inconsistent with classical size–distance invariance, some factor other than perceived distance must be responsible. One of these factors might be perceptual learning.

Berkeley (1709) seems to have been the first author to make a clear statement about the effect of perceptual learning on perceived size. He stated in *A new theory of vision* that we learn to associate certain cues with distance, and also with increased perceived size. He explained the enlarged horizon moon as due to many cues that are normally associated both with increased distance and with increased size. One of these cues was 'faintness' (or aerial perspective – see Chapter 6). Other cues, such as posture, contributed to what is now known as the *angle of regard* (Chapter 11) or to vestibular effects (Chapter 12): they, too, became linked through experience to perceived size. (Berkeley's views on this subject are quoted in detail in Chapter 11.) Without such cues, all objects would be seen in proportion to their angular sizes, and distant objects would appear very small. Berkeley believed in perceptual learning for horizontal viewing, and he explained the small zenith moon as due to a failure of size scaling through inexperience of vertical viewing.

There are many other circumstances besides the celestial illusion in which size constancy (or size–distance invariance) appears to break down. One example is optical magnification (Chapter 5), where the perceived linear size is usually less than that predicted by the perceived distance. A suggestion about a different type of constancy failure was made by Ross, Jenkins, and Johnstone (1980). They proposed that the nervous system has a mechanism for perceiving true angular size independently of any size-constancy scaling, and that this mechanism operates for high spatial frequencies (small angular size), while constancy scaling operates for lower spatial frequencies (large angular size). They claimed that size constancy 'fails' for objects subtending angles of less than half a degree – slightly less than the image size of the sun and moon. As evidence, they claimed that the moon illusion disappears if the moon is diminished by viewing through a reversed telescope. They were apparently unaware that viewing through a reduction tube may reduce or abolish the illusion (Chapter 9). It is hard to understand how constancy scaling could operate for the larger aspects of a visual scene, but not for the fine detail within it – some curious distortions should result. The anecdotal evidence of these authors was disputed by Day, Stuart, and Dickinson (1980), who described experiments showing normal constancy for objects as small as 0.25 degrees. They pointed out that viewing the moon through an optical instrument produces framing effects, and may change the perceived distance rather than the perceived size (Chapter 5). They therefore had their subjects project a small afterimage (0.17 degrees) to the horizon and zenith of a daylight sky. The horizon image was judged significantly larger than the zenith image by a factor of 1.16, showing that some illusion is present for images of small angular size.

A product of size constancy

While the above authors talked about a failure of size constancy under certain conditions, others have described the illusion in terms of normal size constancy and perceptual learning. Reed (1984, 1985, 1989, 1996) proposed a 'terrestrial passage' theory, that relied on *contrast with expected angular size*. He argued that there were two reasons why the moon looks large on the horizon. First, an object with a visual angle of half a degree, at the distance of the horizon or beyond, must be physically large and is perceived as such; furthermore, the moon's visual angle is larger than those of other perceived or remembered (large) horizon objects, causing it to appear angularly and or physically large in comparison. Second, our experience with terrestrial passage (e.g. an aeroplane passing overhead at a constant height) leads us to expect an expansion of an object's visual angle with elevation – an expansion which Reed described mathematically. When seen at high elevation, the moon does not have the large visual angle that we would expect from its remembered visual angle at the horizon. This discrepancy causes us to perceive it as physically smaller and/or further away and/or smaller in angular size.[18]

While this theory is one of perceptual learning, it differs from that of Berkeley. Berkeley relied on the supposed lack of experience when looking upwards, while Reed emphasized the ample experience we have of the behaviour of overhead objects. The arguments are similar to the role of experience in distance judgements (Chapter 9), except that these authors discuss size without appealing to distance as an explanation.

The field of attention

Yet another account of the effects of intervening objects was put forward by the astronomer John Herschel (1833). If terrestrial objects are interposed close to the sun or moon, we judge the latter in the same manner as terrestrial objects, with an acquired habit of attention to detail. Intervening objects do not so much increase the apparent distance of the horizon moon, but instead cause our field of attention to be limited to smaller visual angles.

Various related accounts were given by several authors. Claparède (1906a) believed that the moon appeared enlarged on the horizon because of our greater 'interest' in it. Dadourian (1946) claimed that, when focusing his attention on the horizon sun, it appeared enlarged by a factor of about 2; but when attending to small irregularities on the horizon next to the sun, the latter appeared even more enlarged. In contrast,

Angell (1932) held an 'inhibitory' view: objects seem *smaller* when viewed with an exclusive attention which cuts out the effects of the surroundings, so that if the horizon moon is closely inspected it tends to resume the 'empty' setting it has when elevated.

One of the most detailed 'attention' accounts was produced by Cornish (1935, 1937), whose landscape drawings we described in Chapter 4. He claimed that a restriction of the field of attention leads to an apparent enlargement of the objects within it – rather like the effect of a small framing window. He measured his drawings, and concluded that the enlargement was a factor of about 1.7 for objects on a flat and distant horizon. The horizon sun, he argued, was similarly enlarged.

Theories concerning the field of attention are hard to evaluate as they tend to be inexplicit and only partly testable. More than one idea is involved. There is the idea that 'attention' or 'interest' makes objects appear larger; and there is a second idea concerning the size of the visual field in relation to which the size of an object is judged. There is also the assumption, made by Cornish and others, that the sizes of drawings made at different times reflect differences in perceived size (Chapter 4).

A neurological basis for size scaling

Various suggestions have been made for a neurological basis for size constancy or for other types of size scaling. There is some neurological evidence to support scaling for distance: Dobbins and his colleagues (1998) found that some cells in visual parts of the brain responded best to a combination of angular size and a particular viewing distance. There is also neurological evidence to support pure size relationships: size contrast illusions have been attributed to adaptation of 'spatial frequency' detecting cells, or to other neural interactions in the brain.[19] Such mechanisms may contribute to size constancy at different viewing distances.[20] Some authors attempt to explain size constancy through the enlargement of perceived size for the central part of the visual scene, which is said to occur because the representation of the relevant part of the retinal image covers more cells at later stages of analysis in the brain. Such an idea is based on the anatomical fact of *cortical magnification*, which enhances acuity for central vision. The small central part of the retina (the *fovea*) contains more densely packed cone cells than the surrounding area, and it projects to a relatively larger region of the primary visual cortex. Schwartz (1980) incorporated this idea into his model of size constancy. When an observer fixates a distant object, it forms a small image in central vision, whereas close objects form larger images that spread further into the periphery: the small central image is therefore expanded neurologically relatively more than the larger image. Such a mechanism might contribute marginally to size constancy, but it would not help to explain why the elevated moon appears smaller than the horizon moon when both are viewed with central vision. Other ideas concerning peripheral vision and the moon illusion are discussed in Chapter 11.

Trehub (1991, pp. 242–7) was novel in developing a neurological model – the 'retinoid' model – that could account both for size constancy and the moon illusion. He took the view that size magnification is expensive in neurological terms, because it involves the use of more networks of cells. The human brain husbands its resources by magnifying only the most 'ecologically relevant' parts of the scene – that is objects in the near distance when looking horizontally, and close overhead when looking up. Humans cannot interact (unaided) with celestial objects or with distant terrestrial objects, so the images for such objects can safely be left relatively small. Size constancy is therefore poor for far horizontal distances, and even poorer when looking upwards. The small perceived size of the raised moon occurs because the images of objects in that part of the visual scene are not enlarged much above their retinal representation. The 3D representation of distance is also shrunk vertically in comparison with horizontally, again for the purpose of minimizing neural resources. Distance is computed within the 3D retinoid system, and is represented by 'sheets' of cells called *z-planes*. The extent of size magnification is linked to the distance plane on to which the image is mapped – though not necessarily in the geometrical manner implied by size–distance invariance. Size and distance scaling serve two purposes: they help to preserve spatial constancy despite changes in the retinal image, and they also enhance attention to the most important parts of the visual scene, whilst using neural resources efficiently. This biased mapping of the visual scene onto brain structures is largely the result of human evolution, but it can be further modified by individual experience. Trehub's model also incorporates the effects of the angle of regard (Chapter 11), because the images are rescaled into 'egocentric space' after taking account of gaze direction. Trehub's model is so wide-ranging that it is difficult to classify: it has elements of size scaling, distance scaling, perceptual learning and angle of regard, all combined with levels of explanation at the purposive level and at the neurological level. The details of the neurological model are, however, too complex for this book.

Perceived angular size

Several recent authors have adopted the framework of perceptual size–distance invariance to describe the moon illusion and other errors of size scaling.[21] In McCready's account (1986, 1999) the moon is perceived to change in angular size, and the perceived angular size affects its perceived distance and its perceived linear size. On this view, both angular and linear size are perceived together. The description becomes very complex, as is shown by McCready's diagram for some possible combinations of perceived angular size, perceived distance, and perceived linear size (Fig. 10.8).

A description of the moon illusion in terms of misperceived angular size is logically possible, in the sense that it preserves the geometry of perceptual size–distance invariance. However, it is not in itself an explanation – it remains to be stated what causes the misperception of angular size. McCready suggests that relative size may be a contributing factor, working through some type of neural contrast mechanism. He notes that such an effect is not large enough to account for the whole of the moon illusion, and that some distance cues must also cause changes in perceived angular size. He suggests that such cues evoke conditioned oculomotor minification or magnification (Chapter 7). He also suggests an empirical definition of perceived angular size – that it represents the difference in angular direction between the observer's eye and the two end-points of the target.

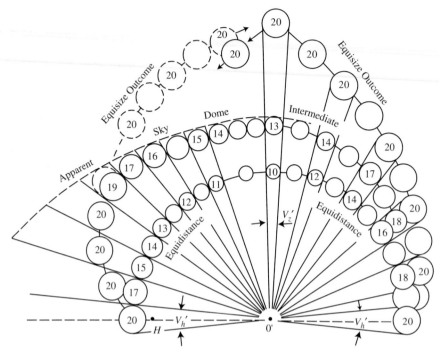

Figure 10.8 Perceived moons with perceived visual angle (V') values that decrease with elevation are shown with six representative examples of the relative [size–distance] illusion that most people suffer. With V'_h arbitrarily chosen to be twice V'_z, the rule $S'_h/S'_z = 2.0\ D'_h/D'_z$ applies. For the equidistance outcomes, D' is 1000 units. For the equisize (size-constancy) outcomes, S' is 20 units. For the three examples in the right half, V' decreases regularly with elevation angle. For the three in the left half, V' decreases more rapidly at first than it does later. The two intermediate outcomes undoubtedly describe the most common moon illusion. (Figure and caption from McCready 1986, with permission of the author and the Psychonomic Society.)

This idea may be relevant for close objects of large angular size, but it is hard to see how it could apply to distant objects subtending only half a degree. The concept of a perceived angular size remains difficult to interpret. We discuss it further in the final chapter.

A hyperbolic scale for perceived angular size

A recent attempt to explain the moon illusion in angular terms has been made by Higashiyama (1992). He showed that observers generally overestimate angular size when asked to make numerical estimates of the angle subtended by an object, or to match the angle by adjustable rods. However, the relation between the perceived and true angles was hyperbolic: small and medium angles were overestimated, but the error diminished for large angles. He used this relation to explain size-constancy scaling, since distant objects subtend smaller angles than near ones and are therefore perceived as relatively enlarged.

Higashiyama went on to argue that the perceived angular size of the moon is equal to the perceived elevation above the horizon of the top of the moon's disc, minus the perceived elevation above the horizon of the bottom of the moon's disc. As the elevation of the moon above the horizon increases, this difference becomes smaller, owing to the negatively accelerated function relating perceived vertical angle to real angle. Hence the perceived size of the moon declines with elevation.

The main problem with this argument lies in the author's strange definition of the angular size of the moon.[22] All reports of the angular size of the moon describe it as a single perceptual quantity, a perceived angular size, rather than the difference between two perceived elevations. Higashiyama's definition is analogous to saying that the (linear) size of a picture on the wall is perceived as the difference between the perceived height above the floor of its top and bottom edges. However, just as the size of a picture can be judged without reference to its height above the floor, so the size of the moon is perceived without first estimating its elevation above the horizon.

The relation between judged and true angular size found by Higashiyama represents a significant contribution to our empirical knowledge about size perception, but its application to the moon illusion remains unconvincing.

The moon illusion and normal size constancy

We have argued that the moon illusion is an enlargement of perceived size near the horizon that is not caused by an enlargement of perceived distance. The enlargement does not seem to resemble local size contrast, but rather a more general assimilation – the moon being perceptually expanded in the same proportion as all horizon objects. We have rejected the idea that the nature of size and distance perception changes as objects rise above the horizon. Instead, differential scaling of perceived size occurs over the whole visual scene. Such scaling is probably achieved through cues such as texture gradients, linear perspective and aerial perspective; and it may also be related to height in the visual scene, with the highest portion being reduced in size.

If size and distance judgements are not highly correlated, it becomes possible to explain why terrestrial horizon objects appear distant, when celestial horizon objects appear close. There are many reliable distance cues – such as interposition, movement parallax, linear perspective, aerial perspective, and texture gradients – which help to create the perception of distance over a terrain and to locate objects within that terrain. The low moon or sun breaks free of these to some extent, so its perceived distance is determined mainly by those cues that affect its perceived angular size.

We have now described several factors that may affect perceived angular size, such as luminance, colour, and relative size. We shall consider some other possible factors relating to viewing angle and bodily posture in the next two chapters.

Summary

Relative size theories are described both for angular size contrast and for assimilation to objects of known size. Simple contrast theories state that the moon is perceptually enlarged in comparison with adjacent objects of very small angular size on the horizon; however, local contrast cannot contribute strongly to the illusion because the illusion occurs in the absence of such objects. Assimilation theories state that the moon is compared with terrestrial objects of known linear size, or is assimilated to the perceived angular expansion of all distant objects. Assimilation theories usually assume that terrestrial objects are scaled for distance; but such size scaling could be achieved by other means. Some authors postulate various types of 'constancy failure' for elevated celestial objects, while others describe the reduction in perceived size as an example of perceptual learning. Some neurological models of size scaling are described. Size contrast is sometimes reported to produce perceptual enlargements of up to 34 per cent, but the effect is usually much less than this. Size contrast may contribute both to size constancy and to the moon illusion, but cannot account for the whole effect.

CHAPTER 11

The angle of regard

In previous chapters we have argued that the moon illusion should be described as a change in the perceived angular size of the moon at different angles of elevation. We have also discussed several factors in the visual scene that might contribute to such changes. We now consider a different type of factor involving a muscular component – raising the eyes in the head. This factor is often confounded with tilting the head or body, which we shall postpone until the next chapter.

The idea that looking upwards diminishes perceived size has a long history. We shall first discuss some early references to the idea, then the experimental literature, and finally some theoretical explanations.

The accounts of Vitruvius and Ptolemy

One of the earliest references to the direction of gaze comes from Vitruvius (first century BC), who stated in his *De Architectura* (III.5 (9)):

> For the higher the glance of the eye rises, it pierces with the more difficulty the denseness of the air; therefore it fails owing to the amount and power of the height, and reports to the senses the assemblage of an uncertain quantity of the modules.[1]

This account has similarities with one given later by Ptolemy in the *Optics* (*c.* 170). Ptolemy's explanations of the moon illusion are confusing, partly because he gave at least two different accounts in different books. In the *Almagest*, his earlier book on astronomy, he ascribed the illusion to some type of refraction by the earth's atmosphere (Chapter 5). Yet in the *Optics* he stated that objects high in the sky seem small because of the unusual conditions and the difficulty of the action. This might be thought to be a forerunner of two modern ideas: perceptual learning, and some version of the angle of regard theory. The relevant passage[2] occurs in *Optics*, III, 59:

> For since it is a universal law that, when the visual ray strikes objects of sight in unnatural and unusual conditions, it has a reduced sensation of all their characteristics, similarly it will have a reduced sensation of the distance which it covers. This would seem to be why, amongst objects in the sky which subtend equal angles between the visual rays, those near the point above our head seem smaller, whereas those which are near the horizon are seen differently and under normal conditions. But objects high in the sky seem small because of the unusual conditions and the difficulty of the action.

This passage is hard to interpret,[3] as it might appear to include two different but partially overlapping explanations: (1) under unnatural conditions, such as looking upwards, the visual ray has reduced sensations of all characteristics (including both size and distance); (2) under unnatural conditions the visual ray has a reduced sensation of distance, which causes a reduction in perceived size (through size–distance invariance).

The visual ray theory (Chapter 3) presents vision as an active process and implies that it can require effort to send out one's rays. For example, Aristotle (*Meteorologica*, III.4) describes how a man with weak sight had difficulty in 'pushing' his visual rays away from him to distant objects – a belief echoed 16 centuries later by John Pecham. The reason for reduced sensations under difficult viewing conditions is probably that it requires more effort to 'push' one's visual rays to the object, and not that some extraneous factor (such as gravity) reduces upward vision.

The other main interpretation of the passage is that overhead objects appear smaller because they appear nearer.[4] It is clear from other passages that Ptolemy was aware of the size–distance invariance principle (Chapter 3). He may also have used the principle to explain the apparent size of planets.[5] Size–distance invariance is the interpretation most commonly ascribed to Ptolemy by modern authors, usually in combination with the intervening objects hypothesis. The latter addition is certainly spurious.[6]

Ptolemy may have given different explanations of the celestial illusion in different books for several reasons: he may simply have been noting all known explanations; his scientific understanding may have improved over the years; or he may not have been primarily interested in the illusion, but needed to account for it when explaining the motions of the planets.[7]

Berkeley's perceptual learning account

The first author to offer a clearly psychological explanation of the angle of regard seems to have been Berkeley (1709). He wrote about the issue in *A new theory of vision*, where he explained the small appearance of the zenith moon as due to the absence of those cues that normally accompany distant horizontal viewing. In particular he mentioned the normal posture of the head and eyes as critical variables. He expounded this view in Section 73:

> The very same visible appearance as to faintness and all other respects, if placed on high, shall not suggest the same magnitude that it would if it were seen at an equal distance, on a level with the eye. The reason whereof is, that we are rarely accustomed to view objects at a great height; our concerns lie among things situated rather before than above us; and accordingly our eyes are not placed on the top of our heads, but in such a position as is most convenient for us to see distant objects standing in our way … It hath been shown, that the judgment we make on the magnitude of a thing, depends not only on the visible appearance alone, but also on divers other circumstances, any one of which being omitted or varied may suffice to make some alteration in our judgment. Hence the circumstance of viewing a distant object in such a situation as is usual, and suits with the ordinary posture of the head and eyes, being omitted, and instead thereof a different situation of the object which requires a different posture of the head taking place, it is not to be wondered at, if

the magnitude be judged different; but it will be demanded, why a high object should constantly appear less than an equidistant low object of the same dimensions ... I answer, that in case the magnitude of distant objects was suggested by the extent of their visible appearance alone, and thought proportional thereto, it is certain they would then be judged much less than now they seem to be, vide Sect. LXXIX. [in which Berkeley asserts that a man born blind who recovers his sight would see all objects in proportion to their retinal image size]. But several circumstances concurring to form the judgment we make on the magnitude of distant objects, by means of which they appear far larger than others, whose visible appearance hath an equal or even greater extension; it follows, that upon the change or omission of any of those circumstances, which are wont to attend the vision of distant objects, and so come to influence the judgments made on their magnitude, they shall proportionably appear less than otherwise they would ... What has been here set forth, seems to me to have no small share in contributing to magnify the appearance of the horizontal moon, and deserves not to be passed over in the explication of it.

Berkeley was propounding a perceptual learning theory (Chapter 9), in which aerial perspective and postural and other cues contributed to perceived size. Without such cues, all objects would be seen in proportion to their angular sizes, and distant objects would appear very small. Our perceptual learning occurs in the horizontal plane rather than the vertical plane; so if the postural cues for horizontal viewing are absent, perceived size is reduced. Berkeley thus believed that there was a failure of size constancy for vertical viewing, due to inexperience of that orientation. The angle of regard was just one of the relevant experiential factors.

Experiments with mirrors

The angle of regard was sometimes considered to be a factor that might explain the moon illusion, and sometimes as one that was itself in need of explanation. Until the nineteenth century, however, there was no observational or experimental evidence to show that it had any effect on perceived size. The German mathematician Karl Friedrich Gauss appears to have been the first to consider how the matter could be investigated. In a letter to the astronomer Friedrich Bessel, written on 9 April 1830, he expressed his dissatisfaction with the intervening objects hypothesis, which was then generally accepted as the explanation of the moon illusion. This hypothesis he considered acceptable only for those people who saw the moon as big as a plate or a cart-wheel; but not for astronomers, who were used to judging only angles on the celestial sphere. Yet astronomers were also subject to the illusion, and this led Gauss to support a physiological rather than a psychological explanation. He then proposed an experiment and reported some personal observations:

> One should perform all sorts of experiments here, e.g. to view the horizon full moon in a plane mirror which reflects it down from some height, without one noticing the mirror or its accessories, and conversely, viewing a reflection of the elevated full moon in a horizontal direction. To satisfy the foregoing conditions sufficiently, such mirrors should however be very large and very accurately flat, and I lack these. However, it appears to me that another experiment indicates a physiological explanation of the phenomenon: when

> I view the elevated full moon from a backwards reclining bodily position, in which the head is in the usual position with respect to the body, so that the moon shines more or less perpendicularly on the face, then I see it much larger; and conversely I see the full moon on the horizon noticeably smaller with the head inclined forwards.[8]

Gauss's proposed mirror experiments, and his informal observations on the effect of bodily posture, form the starting points of two continuing series of experiments. We shall discuss the mirror technique here, and the effects of bodily posture in the next chapter. The experiment with mirrors can be used to test the idea that looking upwards reduces perceived size. If the observer looks up into an elevated mirror to view the reflection of the horizon moon, and the moon still appears as large as when viewed directly, then its normally small size when elevated must be due to factors other than the act of looking upwards. If, however, it appears smaller, the change could be caused by looking upwards, since all other circumstances supposedly remain the same. As with so many other proposed experiments, this one is simple in principle, but difficult in practice. A major difficulty was pointed out by Gauss – which suggests that he actually tried the experiment but was not satisfied with the results. The problem is that an imperfect mirroring surface distorts the moon's image and makes the mirror itself visible: this affects the perceived distance of the mirrored moon and therefore also its perceived (linear) size. This problem has confounded most attempts to put Gauss's proposal into practice.

Helmholtz (1856–66, pp. 290–2) tried a version of the experiment and found that the elevated moon was not enlarged when viewed horizontally in a mirror. However, the experiment does not seem to have impressed him very favourably, for he supplied few details and ascribed the lack of an illusion to the absence of aerial perspective when viewing the mirror image. Stroobant (1884, p. 727) was the next to perform the experiment, and found that the horizon enlargement disappeared when viewing the sun (but not the moon) in an elevated mirror. He too gave very few details.

The difficulties with mirror experiments were first investigated thoroughly by Filehne (1894). He noted that the image of the moon in a mirror usually seems to be very near, and therefore quite small: we are inclined to see the image at the distance of the mirror. Similarly Cave (1938) reported that when the elevated constellations are seen reflected in still water they look 'absurdly small': 'The mind places them on the surface of the water quite close at hand' (p. 290). Another difficulty, mentioned by both Helmholtz and Filehne, is that the reflection of the elevated full moon in a mirror at the same height as the viewer's eye is too bright to fit into a dark horizon scene, and is therefore perceived differently from the genuine horizon moon. Filehne (1894) found that special efforts were required to make a mirror experiment realistic, but that when this was achieved the angle of regard had no effect on the strength of the illusion. These conclusions were confirmed by several investigators a few years later.[9]

Despite these many negative results, it was the later positive findings of Holway and Boring that reached the textbooks and caught the popular imagination. These authors conducted a series of experiments around 1940, with and without mirrors. They measured the perceived (angular) size of the moon at various elevations by matching it to an adjustable circle of light projected on a screen at a distance of 3.5 m. In one experiment

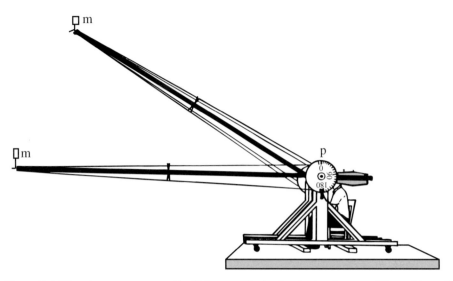

Figure 11.1 The mirror apparatus used by Holway and Boring for viewing the moon. The positions of the mirrors (m) are controlled by strings. Each mirror is at the end of an arm that can be elevated, and the angle of elevation is read on the protractor (p). The observer is shown sitting behind the protractor. The projection screen for an artificial moon is not shown. (Reproduced from Boring 1943 with permission from the *American Journal of Physics*.)

(1940a), the two authors and a third observer[10] viewed mirror images of the moon at various elevations: the observers sat upright and raised their eyes and head as necessary (Fig. 11.1). The matched size of the moon's mirror image decreased as the image was elevated. With the mirror at a distance of about 0.6 m, the angular size of the horizon image was overestimated by a factor of 4.73, and the image raised to 90 degrees was overestimated by a factor of 3.13, giving a ratio of 1.51 for the horizon to zenith moons. At this distance the mirror was visible, but when it was moved to about 6 m it became invisible, and the effects were slightly greater. In a subsequent experiment (1940b), the observer's head could be fixed by means of a bite board and the mirror images were viewed either by turning the eyes only, or by tilting the head only. The two authors served as observers. With head tilt alone, no difference was found between elevations of zero and 30 degrees, the matched size showing an overestimation of 3.36 in both cases. However, when only the eyes were turned, the image in a horizontal direction was overestimated by 3.44 and the image at 30 degrees by 1.84, giving a moon illusion of 1.87. Lowering the mirror below the horizontal showed that turning the eyes down decreased the perceived size of the image to the same extent as raising them – an important finding for explanations of the angle of regard effect. The authors also observed the mirrored sun through a dark filter at various angles of elevation or depression, and made numerical ratio estimates of the perceived size. They obtained similar results to the mirrored moon, but remarked that the sun looked too small through filters and that they found the judgements difficult. They concluded from their mirror experiments, and from other similar experiments in various bodily orientations, that an effect as large as the moon illusion could be obtained purely by raising the eyes in the head. However, they used only two or three observers, and such large effects have not been replicated by other investigators.

Experiments with artificial moons

The use of mirrors proved to be an unsatisfactory technique for measuring a probably small effect of eye elevation. Investigators therefore turned to the study of artificial moon illusions in the laboratory. In most of these experiments an attempt was made to keep the room darkened, so that no change would occur in the visual scene when the eyes were raised.

The first experiment of this nature seems to have been performed by Stroobant (1884). His observers were seated in an empty darkened room so that their eyes were equally far from the wall ahead and from the ceiling. Two small lights, fixed 20 cm apart on the ceiling, indicated the size of an imaginary object. The observer had to adjust the distance between a similar pair of lights on the wall in front of him until the two separations appeared equal. The average result for two observers was that the separation between the lights in a horizontal direction had to be set at about 80 per cent of those on the ceiling in order to appear the same. Similar results were obtained when the separations between pairs of stars were compared close to the horizon and zenith. These results show an artificial moon illusion of 25 per cent – considerably smaller than the natural moon illusion.

Stroobant's results actually contributed little to our understanding of the moon illusion. He did not clearly distinguish between the effects of eye, head, and body movements; it is not known to what extent distance cues were eliminated; and it is unclear whether angular or linear size matches were required. Despite its inevitably inconclusive nature, the experiment was repeated by later investigators. Usually some effect was found[11] but its cause remained unclear. Sometimes no effect was found (Bourdon 1899; Reimann 1902b). In Bourdon's experiment the negative result can be explained by differences in procedure. Using only himself as observer, and with monocular vision, he viewed illuminated circles of light in the dark at various angles of elevation. The upper circle was placed at a distance of 2 m from his eye, at angles of elevation of up to 45 degrees. The lower circle, which had the same diameter of 29.5 mm, was placed in a horizontal location, and could be varied in distance so that its perceived angular size matched that of the higher circle. Bourdon found no significant difference in the distance settings for different angles of elevation. However, an important factor in this experiment was that his head was tilted back at an angle, so that when viewing the 45 degree target he turned his eyes up 20 degrees, and when viewing the horizontal target he turned them down 25 degrees. Since a diminution in perceived size may occur for both raising and lowering the eyes, the upper and lower effects may have equalled each other. Bourdon repeated his experiment with binocular vision, with similar results, but reported that it was very difficult to make judgements of perceived angular size because the changes in distance were too distracting. It is not surprising that he failed to find an 'angle of regard' effect.

Schur (1925) thought that the angle of regard might produce different effects at different viewing distances, and she therefore performed several experiments in various large halls, including a church. She projected circles of light at equal distances on the ceiling and wall in front of the observers (of whom there were between four and seven), and asked them to adjust one of these to match the other in perceived size. The circles appeared larger horizontally, the illusion increasing from a ratio of about 1.15 at a viewing distance of 3 m to 1.88 at 33 m. However, since the observers tilted their heads in addition to turning their eyes, the illusion could not be ascribed to just one of these movements. Also, the halls were not completely dark, so that the illusion was probably affected by distance cues. In further experiments observers were required to move only their eyes to view the upper target at elevations of 25 and 35 degrees. At a distance of 4.8 m there was no illusion at all for six observers, and at larger distances the illusion was still quite small.

Many experiments were designed specifically to test the eye elevation hypothesis. Usually the perceived size of a test object was measured at various angles of elevation, allowing the observers to move only their eyes. Guttmann (1903) judged the separation between two black lines against a uniform background at viewing distances of 25 to 36 cm, with one pair of lines in a horizontal direction being adjusted to match the separation between another pair at an elevation of 40 degrees. With himself as observer, Guttmann found an average illusion of about 3 per cent when the eyes were raised in the head. When he turned his eyes down by 40 degrees the illusion was negligible. He repeated the experiment with another observer, using circles of light as stimuli, and found the same results.

Afterimages were used by some investigators, with variable results. Angell (1924) performed an experiment superficially similar to that of Guttmann, but using afterimages viewed against a background at a distance of 46 cm. An unnamed number of 'experienced subjects' made judgements concerning the fit of the afterimage to an outline drawn on the background. The observers judged whether it fitted or 'spilled over', and made no direct estimates of its perceived size. Angell found that the afterimage fitted the outline, whether or not the eyes were elevated by 40 degrees. However, one would not expect any change in the judged size of the afterimage with such a method, since it resembles the 'marker' or 'bracketing' method discussed in Chapter 8. With this point in mind, Baird, Gulick, and Smith (1962) compared judgements of the size of afterimages both with a 'physical size' (bracketing) method and with a 'perceived size' method. The observers viewed square afterimages at a projection distance of 32 cm, with eye angles of 30 degrees up and down, or horizontal. When using a bracketing method, nine observers showed no difference in judged size with changes in eye elevation. The other method involved adjusting the width of a horizontally located variable square so that it matched the perceived size of the afterimage. A further nine observers used this method. Their settings were significantly smaller by about 3 per cent when the eyes were raised, and slightly (but non-significantly) smaller when the eyes were lowered. The authors noted the breakdown of size–distance invariance, since the projection surface appeared nearer horizontally than in the upper position.

Some other experiments with artificial moons have found almost no effect of eye elevation. Morinaga (1935) tested three observers who viewed white discs of various sizes, with various combinations of head tilt, eye elevation, and disc orientation: he found that the perceived size was greatest when the disc was in the horizontal orientation, eye and head tilt having little effect. Gruber, King, and Link (1963) tested 22 observers who viewed luminous balls in the dark with monocular vision, their heads being fixed so that only the eyes were raised. The observers made ratio judgements of the perceived size of an elevated moon (at about 17 degrees) and a horizontal moon (at about 2 degrees). There was no effect of the angle of regard when viewing totally in the dark, but when a luminous ceiling or other perspective cues were present during or just before making the judgements, an illusion of up to 61 per cent was found. The authors also obtained distance judgements from their observers, but found a low correlation between the size and distance judgements. They nevertheless ascribed the illusion to perspective effects.

An effect of eye elevation was found by Hermans (1954), who devised a more complex artificial moon. He made use of a telestereoscope, with variable convergence angle and angle of eye elevation, to obtain stereoscopic fusion of two illuminated apertures; the perceived size of the resulting circle could then be matched by an adjustable circle viewed horizontally through an ordinary stereoscope. Hermans tested 49 students as observers, and found a decrease in perceived diameter of about 2.4 per cent on raising the eyes 40 degrees, and of about 1.7 per cent on lowering them 40 degrees. He found that both convergence and elevation angles contributed to the diminution in perceived size, but that convergence contributed more. He realized that the size of his effect was much smaller than that reported by Schur or Boring, but put it down to differences in apparatus and to the smaller number of observers tested by those investigators.

Many authors denied any large effect of eye elevation on perceived size, on the basis of everyday experience. Kaufman and Rock (1962a) did so, but they also investigated the matter with the aid of their moon machine (Fig. 4.1). Two artificial moons were projected into the sky, at elevations of zero and 70 degrees. The observers adjusted one of these to match the other in perceived size, using each as a standard on separate trials. Two conditions were compared: tilting the head without lifting the eyes, and raising both head and eyes. Ten observers were tested using both methods, half of them in the reversed test order. The magnitude of the illusion was 46 per cent and 48 per cent respectively for the two conditions, showing a non-significant enhancement of about 2 per cent due to eye elevation. An even more direct test of the eye elevation hypothesis was performed by projecting two artificial moons in the sky at the same elevation of 20 degrees. The ten observers were required to adjust one of these to match the other in perceived size, viewing one with upturned eyes and the other with level eyes and tilted head. There was no significant effect due to raising the eyes, though the mean judged size was about 4 per cent smaller in that condition. The authors also conducted another experiment in a dark planetarium, in which circular patches of light were projected to the ceiling and horizon walls: five observers matched the two against each other, with free eye and head movements, and judged the raised lights to be about 3 per cent smaller than the horizontal lights. Reviewing the subject about 30 years later, Kaufman and Rock (1989) accepted that there might be a very small effect of eye elevation, probably due to convergence.

The work of Kaufman and Rock was highly influential in discrediting Holway and Boring's angle of regard hypothesis, and in re-establishing intervening objects as the accepted explanation. However, they did not succeed in preventing further angle of regard experiments. Bilderback, Taylor, and Thor (1964) tested 90 observers in a darkened room. Artificial moons were displayed at a distance of 2.7 m in both the horizon and zenith positions. A luminous blue silhouette of a city skyline could be interposed in front of the 'moon', and a luminous blue square placed behind it to simulate a sky background. The observers made numerical distance estimates in feet, and they made size estimates by matching the moon to luminous circles provided by the experimenter. There was no significant effect of the presence of the silhouette or background, but there was a significant effect due to position, the horizon moon appearing larger and nearer than the zenith moon. The size estimates gave an angle of regard effect of 6 per cent, and the distance estimates 26 per cent.[12] However, like many other investigators, the authors failed to distinguish between head tilt and eye elevation.

Encouraged by the previous finding, Thor, Winters, and Hoats (1969) tested 128 children of various ages. The children sat in a chair tilted back at 45 degrees, and viewed glowing lamps in a dark room. These lamps were electroluminescent panels of 58 or 24 mm in diameter. They were presented at raised angles of 10 to 90 degrees, at a distance of 2.13 m from the eyes. The children moved only their eyes to view the lamps. The higher lamps were judged smaller and more distant than the lower lamps. There were no age trends, the illusion apparently being fully present by five years of age. Unfortunately, the magnitude of the illusion was not quantified, since the form of the judgements was a forced choice as to which of two lamps appeared further, and which larger. In 1970 the same authors reported a slightly different experiment in which the illusion was quantified. They tested 80 schoolchildren, with a mean age of 14 years. The children used monocular vision, and matched the size of a variable target at one orientation to that of a standard target at the other orientation. The orientations of the variable and standard were counterbalanced, and the lower target was on average judged to be larger than the raised target by about 7 per cent.[13]

Recent laboratory experiments have confirmed an effect of eye elevation. Heuer, Wischmeyer, Brüwer, and Römer (1991) found that raising or lowering the eyes by 30 degrees reduced perceived size by about 5–6 per cent. They tested 24 observers for each condition. The observers made absolute size judgements, using one of six size categories to judge the size of circular rings presented on a monitor screen. The images were presented to each eye separately, at an optical distance of 42 m, and fused with binocular vision. The observers were first trained to make judgements with horizontal gaze, and were then tested with horizontal gaze and with raised or lowered gaze. This method avoided the difficulties of frequent shifts of gaze.

An effect of eye elevation was also confirmed by Suzuki (1991), who tested 16 undergraduates in a planetarium. Two pairs of lights were projected onto the dome screen at a viewing distance of 9.65 m, with elevations of 3 and 72 degrees. The standard pair was always at the higher elevation and had a lateral separation of 62 cm (subtending a visual angle of 3.5 degrees). The separation of the variable pair, at the lower elevation, was adjusted by the observer to appear equal to that of the standard pair. The experiment was conducted either in complete darkness, or with lighting, or with only a horizon and stars visible. Observers either raised their eyes while keeping their neck and body fairly still (though some movement was necessary), or they kept their eyes level while moving their neck and body. Curiously, there was no significant illusion in the lighted room condition; but for the other two conditions the horizontal pair was given too small a separation, showing that it appeared to have a larger separation than the raised pair. The mean size illusion with raised eyes was 33 per cent in darkness and 28 per cent with horizon and stars visible, and the effects were slightly less with neck and body tilted (26 and 16 per cent respectively). These effects are quite large, but it is unlikely that they were entirely due to the angle of regard since they were abolished in the lighted room condition. Failure to counterbalance the positions of the standard and variable targets makes the results hard to interpret.[14] Suzuki (1998) conducted another experiment in a dark planetarium with a viewing distance of 11 m: the raised lights had an elevation of 50 degrees and the comparison lights were viewed horizontally. The mean results for 20 observers with eyes elevated and minimum head tilt gave a size illusion of 37 per cent with binocular vision but only 13 per cent with monocular vision. Suzuki concluded that the effect was largely binocular.

In most of these experiments it is not possible to say whether perceived angular or linear size was measured. However, the size of the effect in percentage terms would be similar. The results obtained in most experiments using large numbers of observers in a darkened room generally show that raising or lowering the eyes (sometimes combined with head tilt) reduces perceived size by about 2–7 per cent in comparison with horizontal viewing. However, Suzuki obtained effects of up to 37 per cent, though the reason is unclear.

Explanations of the angle of regard effect

The term *angle of regard* is sometimes used loosely to mean any effect caused by eye elevation, head tilt, body tilt, or object orientation. Explanations concerned with head and body tilt will be discussed in Chapter 12. Few authors have offered any clear explanations of the effect of raising the eyes, but several attempts have been made. They are described below.

Gravity flattens the eyeball

Schaeberle (1899) thought that gravity had the effect of flattening the eyeball, so that its horizontal diameter would always be larger than its vertical diameter, irrespective of the direction of viewing. Thus when looking towards the horizon, the distance between the lens and the retina would be larger than when turning the eyes upwards. The increase in lens distance would cause a larger retinal image in horizontal viewing, which could explain the moon illusion. This rather peculiar idea may have been inspired by one of the early theories of accommodation, which held that the length of the eye from the lens to the retina varies in order to focus the image on the retina. This theory was tested and disproved in a very courageous experiment performed by Thomas Young around 1800. He inserted a metal hook at the outer angle of his eye when it was turned towards the nose, so that the point of the hook pressed on the middle of the back of the eyeball, causing a bright spot (a phosphene) to be seen. He then changed the accommodation of his eye by focusing it at various distances, and showed that this did not displace the hook or cause any changes in the phosphene.[15] Changes in accommodation therefore do not distort the eyeball, and neither is there any evidence that tilting the eyes in relation to the pull of gravity has any effect on its shape.

The effect of peripheral vision

The part of the retina concerned with acute central vision (known as the *fovea*) occupies about five degrees – considerably greater than the angular size of the sun or moon. The outer area of the retina (the *periphery*) contains fewer retinal cells and has lower visual acuity. Towards the end of the last century Bourdon (1898, p. 394) proposed that the reduced visual acuity for objects in peripheral vision might play a role in the celestial illusion: 'If one draws the curve of the apparent reduction of the size of the moon on leaving the horizon, this curve diminishes rapidly at the start like the well-known diminution of visual acuity between central and peripheral vision' (our translation). The main interest of Bourdon's speculations lies in the fact that they have been consistently misinterpreted. Thus Mayr (1904a, p. 394), Haenel (1909, p. 164), and even Bourdon's fellow countryman Claparède (1906a, p. 125) rejected his proposal on the grounds that the elevated moon appears small even when viewed directly rather than in peripheral vision. Mayr clearly interpreted Bourdon as saying that an object seen peripherally appears smaller – a theory which received some support at that time and is still investigated today.[16] However, Bourdon's idea was different. What he actually proposed was that, when looking directly at the moon, the terrestrial objects between the observer and the horizon become less recognizable as the moon rises above the horizon, because they are seen only out of the corner of the eye. This he thought would cause a change in the moon's perceived distance and thereby also in its perceived size. His explanation is really one of perceived distance, and does not primarily involve factors relating to the elevation of the eye. Dadourian (1946) gave a rather similar explanation based on the visibility of intervening objects, but argued that the illusion was due to comparison with such objects rather than to a change in perceived distance.

Binocular disparity

Binocular disparity (or retinal disparity) is the difference in the images of an object received by the left and right eyes because of their different viewpoints. Fusion of disparate images gives objects 'solidity'. However, the whole field of view cannot be fused at one time, and double images result in front of and behind the point of fixation (see Fig. 7.2). Usually the double images are not consciously perceived, but they do add depth to the scene. The binocular disparity can be *crossed* or *uncrossed*, depending on whether the object is behind or in front of the fixation point. With an uncrossed disparity the left eye sees the left image, and the right eye the right image, while with a crossed disparity the situation is reversed. The brain uses the direction and size of the disparities as a cue to the relative distances of objects in the visual scene. This ability is used to great effect in viewing stereoscopic pictures.

Sir David Brewster (1849/1983), when describing his lenticular stereoscope, pointed out that the same image size appeared either nearer and smaller or further and larger depending on the direction of the disparities. He stated that the moon illusion was similar, in that the moon appeared near and small in the zenith and distant and large on the horizon. He did not, however, claim that the moon illusion was caused by binocular disparities, but merely by a change in perceived distance. Lewis (1862) made a similar point, giving a change in disparity as an example of a factor producing a perceptual size effect large enough to account for the moon illusion. Lewis did not use a stereoscope for his demonstration. Instead, he used the method of voluntary changes in convergence to fuse a pair of stereoscopic pictures of the moon, fixating either behind or in front of the pictures. The diagram that Lewis gave is reproduced in Fig. 11.2. Near fixation of course introduces oculomotor micropsia (Chapter 7) in addition to the crossed retinal disparities, so it is not clear which of these factors is responsible for the change in perceived size.

Unfortunately, Lewis did not publish his stereo pair of moons. However, the reader may like to explore Lewis's effects with the off-centre double circles shown in the middle of Fig. 11.3. The left and right figures are reversed in the lower pair. If the upper and lower figures are fused with the same vergence of the eyes, the binocular disparities differ in direction and one of the fused circles will appear behind the page and the other in

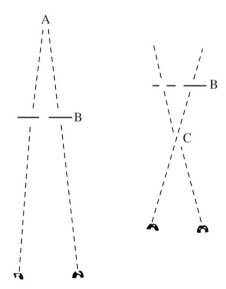

Figure 11.2 The figures from Lewis (1862) to illustrate crossed and uncrossed fusion obtained by changes in fixation. The two stereoscopic pictures of the moon were located at B. When Lewis fixated behind the pictures he saw a solid spherical moon located at A. When he fixated in front of the pictures he saw a small concave ('inside out') moon located at C.

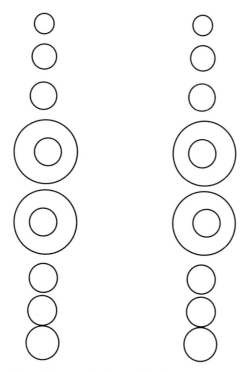

Figure 11.3 The inner circles of the central pairs may be fused by relaxing fixation behind the page. These circles are all of equal size on the page. However, when fused, the upper circle appears nearer and smaller than the lower circle. These effects are reversed if fusion is obtained by converging the eyes in front of the page. The difference in perceived size can be measured by matching each of the fused inner circles to one of the upper or lower reference circles. If the inner circles have a diameter of 1.0, the three upper circles have diameters of 0.76, 0.92, and 1.00, and the lower circles 1.04, 1.11, and 1.21. (The central display was copied from Sanford 1898, p. 276, and the reference circles were added by Lloyd Kaufman. With permission of Lloyd Kaufman.)

front. Fusion can most easily be achieved by holding a card vertically down the centre of the two pairs and placing the tip of the nose at the top of the card: this allows the two eyes to become nearly parallel, or to converge beyond the plane of the paper. With this method of fusion the upper circle appears to be close and small and the lower circle further and larger: the difference in perceived size is about 20 per cent for many observers, which would be too small to account for the moon illusion. (The reader can check the size of the effect by matching the perceived size of each of the fused central circles with one of the circles ranged above or below: the difference between the matched sizes gives a measure of the effect.) An alternative method of fusion is to hold a pencil in front of the diagram and converge the eyes on the pencil. With this method the effects for the upper and lower circles are reversed. The relative difference in perceived size between the upper and lower circles remains the same as with the relaxed vergence, but the whole display appears strikingly smaller with strong convergence.[17] This latter effect demonstrates the strength of oculomotor micropsia: it is perhaps a size change of

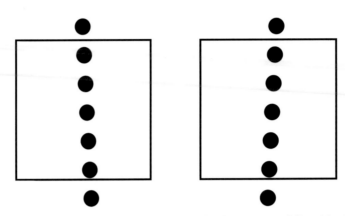

Figure 11.4 The circles are all of equal size on the page, but have a range of disparities. They can be fused to give a series of stereoscopic 'moons' of increasing or decreasing perceived distance and size. (Display by Lloyd Kaufman, with permission.)

50 per cent, which (if proved to be relevant) would be sufficient to account for the moon illusion. A whole range of stereoscopic 'moons' of different disparities is shown in Fig. 11.4, and again the depth and size effects can be reversed with near or far fixation.

Binocular disparity was invoked by Holway and Boring (1940b), in a theory which they attributed to Adelbert Ames of the Dartmouth Eye Institute. They suggested that retinal disparity should increase with a raised angle of regard, because raising the eyes causes them to rotate slightly outwards (excyclotorsion). The brain might take the increased disparity as equivalent to that produced by objects nearer than the fixated object. Increased disparity should occur in the upper field of vision when the eyes are raised (owing to excyclotorsion) and in the lower field when the eyes are lowered (owing to incyclotorsion). Thus both raising and lowering the eyes would provide a cue to reduced distance for objects just above or just below the point of fixation. This cue, together with the size–distance invariance principle, could account for a reduction in the perceived linear size of objects seen when looking either up or down.

Holway and Boring were dissatisfied with this explanation because the objections mentioned earlier still applied: the horizon moon tends to appear nearer instead of further than the zenith moon, and the illusion is seen by monocular people. They did not mention another obvious objection – that one would have to be fixating below the moon and not directly at it in order to perceive such an effect.

Retinal illumination

Holway and Boring also discussed a reduction in illumination as a possible cause (see Chapter 6). They argued that raising the eyes should reduce the illumination on the retina because of partial cover by the eyelid, and because of the oblique angle of entry of the light rays through the pupil. However, they concluded that such an effect would be too small to account for the moon illusion.

Convergence effort

Several theorists followed Berkeley in supposing that motor activities, such as changing the orientation of the head and eyes, become associated with changes in perceived size. One version of this idea was the convergence effort theory. It was proposed by Stroobant, and expounded in more detail by Zoth (1899, 1902). Referring to the work of the physiologist Ewald Hering, Zoth argued that when the eyes are turned upwards they also tend to turn slightly outwards (excyclotorsion), as a result of the position of the muscles controlling the movement. This purely mechanical effect also operates when viewing with one eye only. Zoth proposed that the divergence is counteracted by nerve impulses causing convergence of the eyes. These convergence impulses become associated with close distances and with a reduction in perceived size, and cause the elevated moon to appear either smaller or closer. When no visual distance cues are present, the elevated moon will therefore appear smaller than the horizon moon, but when distance cues are present, it may look either smaller or closer.

Zoth's explanation has some similarities with the motor command account of oculo-motor micropsia discussed in Chapter 7. However, his argument is unclear because it confuses angular and linear size. The argument could take two forms: first, that convergence effort acts as a cue to close distance, and a reduction in perceived linear size follows from size–distance invariance; or that convergence effort acts to reduce perceived angular size, independently of perceived distance. The second account is preferable, since observers often report that objects appear both small and distant.

Boring was unable to find an acceptable explanation of the moon illusion. He was inclined to explain size constancy in terms of classical size–distance invariance, but realized that the moon illusion was a size–distance paradox. He was therefore looking for some explanation in terms of the angle of regard. He wrote 'Now you want to know *why* raising the eyes shrinks the moon, but no one has yet been clever enough to formulate for test the crucial hypothesis that will answer the question' (1948, p. 16). Holway and Boring (1940b) discussed and rejected several theories. They rejected the convergence effort theory, because they thought convergence effort should act as a cue to near distance, with reduced perceived (linear) size caused by size–distance invariance; yet the zenith moon appears further and not nearer than the horizon moon. Moreover, the illusion is seen by one-eyed people, whom the authors thought should not have a convergence tendency. The convergence theory would also predict that objects seen with lowered eyes should appear larger, because a compensatory divergence occurs which should be a cue to increased viewing distance. However, objects appear reduced rather than enlarged in size when looking down, which suggests that the convergence explanation is incorrect. Holway and Boring went on to discuss an explanation related to binocular disparity, which we described earlier. Finally they returned to the idea of Zoth and others that the effort of raising or lowering the eyes somehow diminishes perceived size, but they remained at a loss for an explanation. 'What we should like to find is some means whereby eye movement above and below the primary position diminishes the total excitation of the visual perceptual field' (1940b p. 553). They thus seem to have been looking for a theory in which perceived size is related to 'total excita-tion', the latter being reduced or 'inhibited' by vertical eye movements. Such an account would seem to be very far removed from either size–distance invariance, or Berkeley's motor association theory. It also bears little relation to any modern ideas on the neural coding of size.

Oculomotor effects were supported by Enright (1989a, 1989b), after he had revoked his 1975 accommodation account (which we described in Chapter 7). He supported Zoth's idea that there is a tendency to diverge the eyes when looking up, citing the work of Holland (1958) who showed that there was about half a degree of divergence for a 15 degree uptilt of the eyes. There would thus be a compensatory effort to converge, causing oculomotor microspia. Enright reported that a 1 degree change in vergence caused about a 30 per cent change in perceived size, but that transient vertical eye movements might produce greater effects. He further suggested that pictorial cues, or vergence related to perspective, would enhance the perceived size of the horizon moon. These and some other effects would be sufficient to account for the full size of the moon illusion.

While convergence effort could play some role in causing microsmia when the eyes are raised, the evidence in its favour is poor. A strong negative argument, mentioned by several authors, is the fact that microsmia also occurs when the eyes are lowered. Divergence effort is needed when the eyes are lowered, so macropsia should occur rather than microsmia. Experimental evidence was also produced by Heuer and colleagues (1991). They obtained measures relating to vergence effort in their observers, and found no correlation between that and perceived size. When vergence effort was constant, perceived size varied with the true angle of convergence: minification of less than about 10 per cent occurred as convergence increased.

Oculomotor conditioning

Most experiments have shown that raising the eyes in the head produces only a small reduction in perceived size. However, other oculomotor factors (such as accommoda-tion and convergence) are reported to produce a much larger effect. We described several experiments on accommodation in Chapter 7, and concluded that the causal relationships were difficult to disentangle. Most authors have stressed the contribution of convergence, and many have given a motor conditioning account similar to that of Berkeley. For example, Hermans (1954) argued that eye movements of elevation and outwards rotation become conditioned to produce a reduction in perceived size, and that these movements could account for the reduced perceived size of the moon when raising the eyes. Heuer and colleagues (1991) made a similar argument for the conditioning of gaze angle.

McCready (1965, 1985, 1986) has offered some different arguments involving oculo-motor microsmia.[18] The various distance cues in a scene evoke conditioned changes in the motor command signals for accommodation and convergence, causing perceived angular size enlargement for distant objects. The conditioned changes may occur whether or not they are appropriate – for example in photographs, certain geometrical illusions, and very distant targets. The isolated zenith moon does not evoke such changes, while the distant horizon scene does so, thus causing the moon illusion. In his most recent statement (1999) McCready argues that the relative enlargement of the horizon moon is the result of two opposite oculomotor illusions: macropsia when looking to the horizon, and micropsia when looking up. When viewing the horizon moon, the details of the landscape provide cues to a great distance, causing the eyes to focus and converge on too far a distance, and inducing oculomotor macropsia that affects all objects in the scene, including the moon. When viewing the elevated moon no distance cues are present, and the eyes tend to adjust to too close a distance, inducing oculomotor micropsia.

One difficulty with most accounts based on oculomotor micropsia is that they imply that the raised moon is perceived as 'too small'. In fact, however, the measurements of Boring and others show that the raised moon is perceived as larger than a close comparison target of the same angular size. Micropsia is a relative term, and is a useful description only for the relation between targets viewed at the same distance but with different oculomotor adjustments.

We can conclude that raising or lowering the eyes reduces perceived (angular) size by about 2 to 7 per cent compared to horizontal viewing. There may also be a small effect due to head tilt or other changes in bodily position, a question to which we shall return in the next chapter. As for its explanation, we need one that will work for both raising and lowering the eyes in relation to the horizon. Oculomotor micropsia relating to convergence effort could work only for raising the eyes. Some type of motor conditioning theory seems to be necessary, in which size reduction is conditioned to looking up or down. Such conditioning would follow from the fact that objects above our head or at our feet are usually closer and subtend a larger angle than those viewed horizontally: we therefore learn to reduce the perceived angular size of objects in these locations, relative to those on the horizon, in order to assist size constancy.

Summary

The obscure views of Vitruvius and Ptolemy on the difficulty of looking upwards are discussed, and Berkeley's views on perceptual learning and the angle of regard are outlined. Experiments on eye elevation using mirrors, artificial moons, and afterimages are described. Many experiments gave conflicting results because they used only small numbers of observers, confounded eye elevation with bodily orientation, varied the presence of visual cues, failed to counterbalance the orientation of the standard and variable targets, and were unclear as to whether perceived angular or linear size was measured. The better controlled experiments show that objects appear both smaller and further with raised or lowered eyes than when viewed horizontally. The perceived size change is typically about 2–7 per cent. Various oculomotor explanations are described. Convergence effort is rejected, because it could apply only for upward gaze. Conditioned oculomotor micropsia for both upward and downward gaze is a preferable explanation.

....

CHAPTER 12

A question of balance

Most of the explanations of the moon illusion discussed so far have a lengthy history, having been around in embryonic form for centuries. These explanations have been concerned primarily with the nature of the *visual* information, and the way the observer interprets that information. The idea that there might be some *sensorimotor* interaction in the illusion is relatively new, and probably goes back to Berkeley (1709). He argued that both size and distance perception are ultimately based on eye movements and tactile exploration; he started a tradition of 'motor theorizing' that reached its peak in the second half of the nineteenth century, and was pursued by experimental scientists such as Lotze, Helmholtz, and Wundt.[1] The sensory aspect of sensorimotor systems is usually called *proprioception*.[2] The proprioceptive mechanisms that contribute to the observer's knowledge of his own bodily orientation include the vestibular system (the balance organs of the inner ear), the pressure receptors in the skin, and the receptors in the muscles, tendons, and joints. The idea to be considered in this chapter is that proprioceptive information interacts in some way with the visual perception of size and distance, with the result that changes in the observer's bodily orientation contribute to the moon illusion.

The relatively late inclusion of proprioceptive explanations for the moon illusion probably stems from the fact that the understanding of sensorimotor mechanisms lagged two or three hundred years behind that of vision. It has always been evident that the eye mediates vision. The eye is a relatively accessible organ, and a good understanding of its mechanism was achieved by the time of Kepler (1611). It was also evident that the ear mediated hearing, but the organ was less accessible for experimentation than the eye, and the physics of sound was less well understood than that of light; consequently, the workings of the ear were not well understood until the time of Duverney (1683). It was not, however, at all evident that the sense of balance was mediated by the vestibular system of the inner ear.

The vestibular system consists of two main parts: the semicircular canals and the otolithic organs. The canals respond mainly to angular acceleration (rotary movement). The otolithic organs consist of the utricle and saccule, and they respond mainly to linear acceleration. Linear acceleration is caused by changes in speed of bodily movement in one direction. However, the linear acceleration of earth's gravity also provides a constant force (known as 1 g). The otolithic organs respond to this force, and so signal the static orientation of the head in relation to that of gravity. It was known since at least the time of Lucretius (*c.* 56 BC) that vertigo could be caused by motion,[3] but the mechanism was not linked to the semicircular canals until the work of Flourens between 1824 and 1830. The functioning of the canals and otolithic organs was explored in more detail by Mach, Breuer, and Brown between 1873 to 1875.[4] It is the otolithic organs that are of prime interest in relation to the moon illusion, because they contribute to knowledge of head tilt and gaze direction.

The muscle, joint, and skin receptors are more visible and open to investigation than the vestibular system. The discovery of the muscle spindles goes back to W. Kühne in 1863, but the other receptors have been described only since the 1880s. The neck joint receptors are again of special interest for the moon illusion, because they also contribute to knowledge of head tilt and gaze direction.

Proprioceptive explanations of the moon illusion

The first proprioceptive theory of the celestial illusion may have been suggested by Ptolemy in his *Optics,* where he spoke of the difficulty of looking upwards. However, as discussed in Chapter 11, it is more likely that he supposed the difficulty to lie in the visual rays rather than in the perceptual system. A much clearer account was given by Berkeley (1709), who suggested that head and eye orientation contributed to size and distance perception through learning. His theory was deficient, however, in that he believed that size constancy was conditioned only to horizontal viewing; upwards viewing he regarded as a case of constancy failure (Chapter 10).

In more recent times, Schur (1925, p. 75) was probably the first to suggest that vestibular and neck joint stimulation might add to the effect of raising the eyes in the head. The first authors to propose an explicit vestibular mechanism in the moon illusion seem to have been Wood, Zinkus, and Mountjoy (1968), who entitled their paper 'The vestibular hypothesis of the moon illusion'. Like many previous authors, they thought that the orientation of the head in relation to the direction of the force of gravity might be important, and they speculated on possible neural mechanisms.[5] Many experiments followed.

Experiments on head or body tilt

It has often been noticed that inverting the head makes objects look smaller and reduces the horizon enlargement of the moon. Some authors have put this down to proprioceptive effects, and others to changes in the visual scene when inverted. Sanford (1898, pp. 211–12) commented that colours appear distorted, and that distances may appear either enhanced or reduced. Some authors have commented on the resulting size–distance paradox. For example, Angell (1932) remarked: 'More pronounced is the effect of gazing with head inverted; objects appear decidedly smaller and likewise farther away, the latter probably a "secondary illusion" which connects smaller size with greater distance.' The effect was also reported by Meili (1960), who noted that the perceived enlargement of distant hills was much less when tilting the head or looking through one's legs: he argued that this was because size constancy depended on the presence of a good *gestalt*, or well-structured visual scene, and tilting the head somehow changed this. Coren (1992) made a rather similar point: he argued that an inverted pictorial scene reduces the 'salience' of depth cues, and depth cues rather than proprioceptive factors were the cause of the moon illusion. Explanations involving an inverted visual scene have even crept into modern popular literature, as in a novel by Atkinson (1997) (quoted in Chapter 13). Boring (1943) also commented that the horizon moon was much diminished when viewed through one's legs, and ascribed this to changes in the angle of regard or the vestibular system. We shall here consider some more systematic experiments relating to bodily posture and proprioceptive factors.

In the last chapter we described several experiments that were principally concerned with eye elevation, but some of which also entailed head or body tilt. It is clear that several factors could be involved. Possible candidates are the extent to which the eyes are turned upwards in the head, the extent to which the head is tilted back with respect to the axis of the body, the position of the head and eyes in relation to the earth's gravitational pull, and the occurrence of sudden head movements causing additional stimulation of the vestibular system.

The effect of various bodily postures was first investigated by Gauss (1880). Gauss found that when his body was tilted backwards so that the elevated moon shone in his face, the moon appeared larger than when he looked up by tilting his head and turning his eyes up. Thus movements of the head and eyes seem to be implicated, though Gauss's limited observations cannot be conclusive. Later investigators have repeated Gauss's experiment rather more formally, with several observers. Reimann (1902b, 1904) made numerous estimates of the sizes of the sun and moon from various bodily positions, but neither he nor any of his observers found any effects of posture. However, Zoth (1902), using 50 observers, confirmed Gauss's results.

Holway and Boring (1940a) used both supine and upright postures when investigating the effects of eye elevation. In the supine posture the observers lay on their backs and tilted their heads so that they were almost inverted. In this position the low moon appeared much smaller than when viewed from an upright posture. However, only two observers took part in both postures, because Boring himself had problems with the supine posture: 'At neither time was E. G. B. able to assume this position without nausea, so that observations from him in the supine position had to be abandoned.' The authors nevertheless concluded that the differences in perceived size were due to eye elevation rather than bodily posture.

Many postural experiments were conducted during the 1960s, but none succeeded in separating all the contributing factors. We shall describe a few of these in detail, to give a flavour of the procedures. The most entertaining were some Chinese experiments involving the use of red balloons. Ching, Peng, and Fang (1963a) required about 48 observers to match the perceived size of an adjustable balloon at a distance of 1 m to that of balloons of 38 cm diameter set at distances from 50–250 m along the horizontal. The instructions for perceived size were to use a 'naive attitude'. Head posture was varied rather than the angle of the eyes in the head. With normal posture, the mean size matches reduced systematically from 32 cm at a distance of 50 m to 24 cm at 250 m. The matched size of the distant balloons was greatest for observers in a normal sitting posture, less for those lying prone or supine and raising their heads to view the balloons, and least for those lying supine on a dais and hanging their head backwards over its edge. This remarkable position also had the effect of increasing perceived distance. The authors concluded that size constancy was more reduced the more unusual the posture. In another experiment (Ching, Peng, Fang, and Lin, 1963b), the perceived size of a 38 cm hydrogen-filled balloon at various angles of elevation, and various distances of up to 250 m, was compared to discs set at 2 m. Up to 43 observers were tested, and the matched size decreased gradually with the angle of regard when observers raised their head and eyes to view the balloon; it also decreased, but more abruptly, when they viewed the elevated balloon straight ahead from a backward tilting chair. This experiment must have been spectacular, and the authors' sketch is reproduced in Fig. 12.1. They also found that when the observer looked down from a high building to the standard balloon on the ground, the matched size diminished, but to a lesser degree than with a raised angle of regard. These experiments are similar to those of Holway and Boring (Chapter 11) – and subject to the same methodological problems – but produced results that could be interpreted in terms of a postural or vestibular effect instead of that of raising the eyes. However, the authors thought that the loss of visual ground cues in unusual postures was an important factor.

Some other experimenters have made use of afterimages. King and Gruber (1962) found that the perceived size of square afterimages depended on their projected position in the sky, or perhaps on head tilt. Their observers viewed afterimages monocularly, tilting the head rather than the eyes, and estimated the ratio of the perceived lengths for horizon and zenith viewing. The average ratio for 14 observers was 1.63. This ratio is almost large enough to account for the natural moon illusion, but confounds the effects of head tilt and the visual scene.

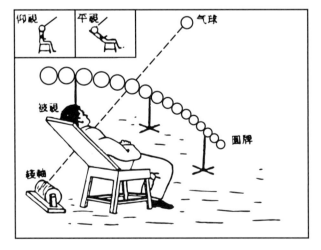

Figure 12.1 The balloon experiment of Ching, Peng, Fang, and Lin (1963b). Permission sought from *Acta Psychologica Sinica.*

Yet another method involved luminous objects viewed in the dark. Bilderback, Taylor, and Thor (1964) performed an additional experiment to the one described in the previous chapter. They attempted to control for eye position (and presumably for neck tilt) by placing the observer in an adjustable chair. The targets were luminous balls located 1.8 m from the eyes. The 27 observers were asked to estimate the ball's distance from their eyes. The mean estimated distance for horizontal stimuli was 2.3 m, which was significantly shorter than the overhead estimate of 2.7 m. Unfortunately no size estimates were obtained. The distance estimates give a distance illusion of 17 per cent; but judging from the authors' earlier experiment the equivalent size illusion would be less than 6 per cent. This result points to a small vestibular or postural/tactile component, though the authors did not specify this and referred instead to Holway and Boring's 'angle of regard' hypothesis.

Luminous circles in the dark were also used by Van de Geer and Zwaan (1964), who performed experiments that combined head tilt and eye elevation, whilst varying the orientation of the body and target. With upright posture their 26 observers made matches showing that the horizontal target was perceived as about 10 per cent larger than the vertical target, but when supine they made accurate matches.[6] The authors concluded that an angle of regard effect was present with upright posture but not with supine posture. In a second experiment, six observers made size matches between targets at different distances: the perceived enlargement with distance was greatest for horizontally located targets and upright bodily orientation, less for an elevated regard of 60 degrees, and least for targets elevated at 60 degrees.[7] The authors argued that size constancy was conditioned to normal viewing, and was less effective in other orientations. A third experiment examined the use of one or both eyes. With upright posture and binocular vision, the four observers judged the horizontal target as 24 per cent larger than the raised target, but with monocular vision only 18 per cent larger. With supine posture, the matches were always correct.[8] Thus binocular vision enhances the angle of regard effect.

Real objects in daylight were used by Ross (1965). She investigated similar variables to those of Van de Geer and Zwaan, and found rather similar results. Her five observers manipulated the distance of a larger circle to appear equal in phenomenal (angular) size to a smaller circle.[9] The furthest distance settings (or greatest overestimation of angular size) occurred for horizontally located targets and upright body orientation. The largest difference in settings (about 21 per cent) was due to the orientation of the apparatus; but the use of an illuminated room makes it likely that differences in the visual background were the main cause.[10] There also appeared to be a small difference of about 6 per cent due to bodily orientation, but none to head/eye tilt.

A vestibular contribution to the moon illusion was explicitly suggested by Wood, Zinkus, and Mountjoy (1968), but they actually measured only the combined effect of head and neck tilt and eye elevation. Their 30 observers were seated in a darkened room and matched the perceived (angular) size of a standard illuminated disc 1.8 m above the chair with a variable disc presented horizontally at the same distance. The horizontal discs appeared larger than the vertical discs by 14 per cent. This is hardly large enough to explain the moon illusion, but it does suggest that the orientation of the head and eyes is a contributory factor. Wood and his colleagues also asked 33 observers to adjust the height of a vertically viewed disc so that it appeared equal in distance to a horizontally viewed disc of equal size. The vertical settings were about 27 per cent closer, showing that vertical discs appeared further away than horizontal discs. The authors interpreted this to mean that the vertical discs were perceived as smaller than the horizontal discs: presumably they thought that the observers were using perceived (angular) size as a cue to distance, and were reducing the distance of the vertical discs so as to increase their angular size. Van Eyl (1968) performed a rather similar experiment involving distance matches, with similar results.

Zinkus and Mountjoy (1969) aimed to control head, body, and eye tilt more precisely. They tested 20 observers, who were seated in a tilting chair with a head rest. The chair was enclosed on the sides and top, apart from apertures for viewing the illuminated stimulus discs in an otherwise dark room. The observer could see the front disc by looking forward, but had to tilt the chair backwards to see the overhead disc, or swivel 90 degrees to the right to see the other horizontal disc. The front disc was placed at distances of 1–2 m from the eyes, and the observer was required to set the upright or right horizontal discs to the same perceived distance as the front disc. The vertical distance appeared greater by about 27 per cent, which suggests that the distance effects found in their previous experiment were due to otolithic or other postural stimulation rather than neck tilt or eye elevation. The authors again interpreted the distance effects as due to a change in perceived (angular) size, through vestibular stimulation.[11]

Another attempt to isolate a vestibular contribution to the moon illusion was made by Carter (1977). He pointed out that the illusion is much reduced with monocular viewing, but that it ought to exist if there were a vestibular component. Like several previous authors, Carter measured only perceived distance and not perceived size. The 24 observers viewed circular targets (screwdriver handles) located in translucent plastic tubes, which were oriented horizontally or tilted 45 degrees up or down. The observers used monocular vision, and tilted their heads to look through the tilted tubes. They had to judge whether the horizontal target (which was variable in distance) was further or closer than a tilted target at a standard distance. The horizontal target was generally judged closer, which means that the targets appeared further away when viewed obliquely up or down. The effect was more marked for downward than upward viewing. Unfortunately Carter did not calculate the point of subjective equality for the distances, so it is not possible to quantify the change in perceived distance. One can conclude only that head tilt contributes to the more distant appearance of objects, even with monocular vision.

The monocular effect was also found to be small by Suzuki (1998). He tested 20 observers in a dark planetarium with a 50 degree backwards head tilt (as part of the eye elevation experiment described in the previous chapter). He found a size diminution of 22 per cent for the head raised with binocular vision, but only 11 per cent with monocular vision. He concluded that the head tilt effect was largely dependent on binocular vision.

Few of the experiments described in this section are sufficiently clearcut to give a reliable estimate of the diminution in perceived size caused purely by head or body tilt. Some authors confounded head tilt with eye elevation, or with visual cues to object orientation. Many authors failed to obtain direct measures of perceived size at different orientations, and instead measured changes in perceived distance confounded with changes in perceived angular size. Nevertheless, several experiments do provide evidence for a vestibular–postural influence on perceived distance, with vertical distances appearing enlarged by up to 27 per cent. The corresponding reduction in perceived size cannot be calculated from this information, because classical size–distance invariance does not hold; however, the experiments all point to some reduction in perceived angular size. The size reduction is probably less than the distance elongation. The only experiments that give a direct measure of perceived size in dark conditions are those of Suzuki (1998), who found a 22 per cent reduction for backwards head tilt with binocular vision; and Van de Geer and Zwaan (1964) who found a 10 per cent reduction for combined head and eye tilt, and perhaps an 18–24 per cent reduction for a supine body position. Unfortunately neither authors counterbalanced the locations of the standard and variable targets, making the true size of their effects uncertain.

Most authors agree that binocular effects are larger than monocular effects for all postural combinations. This may point to the role of convergence. Heuer and Owens (1989) showed that forward and backward head tilt had effects on resting vergence similar to those of lowering and raising the eyes, though rather smaller. As was pointed out in the last chapter, this finding is contrary to the convergence effort theory, because lowered eyes and forward head tilt should make objects appear larger rather than smaller.

Vestibular experiments using applied accelerations

The best way of stimulating the otolithic system independently of the joint receptors in the neck is by tilting the whole body (including the head) on a tilt table or suitable chair. In that case the otoliths are stimulated by the accelerative force of earth's gravity – defined as 1 g in the z axis. We have described some experiments of that type in the previous section. It is also possible to stimulate the vestibular system through various types of applied acceleration, such as road or rail travel, the parallel swing, the human centrifuge, air travel, and space travel. Some of these methods stimulate the otolithic system in a way that mimics natural head tilt. The methods are, of course, more expensive than using a tilt table, and they often introduce unwanted complications. Nevertheless they have been used over the last half-century to investigate vestibular effects on the moon illusion, or, more generally, on size and distance perception. Many of the experiments described here were connected with aviation or space research rather than the moon illusion.

In the descriptions that follow it is sometimes necessary to refer to the x, y, and z axes of the human head or body, which are shown in Fig. 12.2. The force of gravity always operates through the z axis of a body that is upright with respect to the earth, whereas

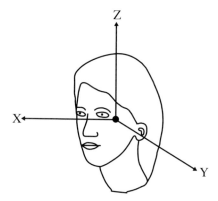

Figure 12.2 The principal axes in relation to an upright head.

the force applied to front- or back-seated passengers in linear travel in road and rail vehicles is mostly in the x axis.

Some of the earliest vestibular experiments were conducted in the human centrifuge, probably because such centrifuges were available for aviation research. The human centrifuge consists of a rotating arm with a cabin at its outer end. The cabin is usually free to swing out in line with the gravitoinertial force (determined by the length of the arm and the angular speed of rotation). When the cabin rotates at a constant angular velocity, it causes a constant linear acceleration: the passenger then feels no tilting force but merely an increase in his body weight. However, when the centrifuge starts up or slows down, the varying angular velocity and varying linear acceleration cause tilting or tumbling sensations, the effect depending on the passenger's bodily orientation within the cabin. Gregory, Wallace, and Campbell (1959) experimented with afterimages, which four observers viewed in a dark cabin. They noted that, during the 'tumbling' phase, caused by deceleration, 'the after-image ... may expand, as though the observer is falling into it'. This effect is hard to explain on the basis of Emmert's law, which states that an afterimage shrinks when the observer approaches the surface onto which it is projected (Chapter 8).[12] However, four observers is too small a number to determine the typical perceptual size change, since large individual variations occur.[13] If head tilt simulated in a centrifuge does indeed cause perceived expansion of afterimages, it would not help to explain the diminution of perceived size that occurs with natural head tilt.

More predictable linear accelerations are produced by the parallel swing. When an observer is seated on such a swing with head erect, he is subjected to a changing linear acceleration in the x axis (Fig. 12.3). There is also a small z axis component due to the vertical motion of the swing, but this can be ignored. Forwards linear acceleration of the swing should stimulate the otoliths in the same way as does gravity when the head is tilted backwards.[14]

Gregory and Ross (1964a) seem to have been the first to investigate size perception with this method of linear acceleration – the intention being to produce results of interest to the US air force. They tested 19 observers who were seated, with their heads

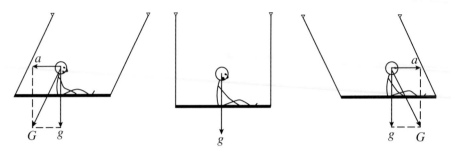

Figure 12.3 The parallel swing used to stimulate the otolith organs sinusoidally. *G* is the resultant of the applied acceleration, *a*, and the acceleration due to gravity, *g*. (After Jonkees 1975.)

supported, on a parallel swing in the dark.[15] The observers made monocular judgements of the change in perceived size of a luminous circle which was displayed on an oscilloscope, located just in front of the frame of the swing. The circle expanded as the swing moved backwards and contracted as it moved forward, thus partly counteracting the change in the observer's retinal image. The rate of change could be varied by the experimenter, and the rate at which no size change was perceived was taken as a measure of size constancy. Size constancy was significantly greater for forwards than for backwards movement.[16] While the results may suggest a vestibular contribution to size scaling during motion, it is not clear what they would imply for an equivalent degree of static head tilt.

A parallel swing was also used by Van Eyl (1972), who was interested in the effect of vestibular stimulation on the artificial moon illusion.[17] Three observers used monocular vision to view briefly presented luminous discs in the dark, both when stationary on the swing and when moving. They compared the relative sizes of horizontal discs with equidistant raised discs. The higher discs always appeared smaller: the maximum illusion was about 8 per cent for the combined effect of eye elevation and head tilt for an 80 degree display. There was no difference between stationary and swinging judgements, nor between the maximum acceleration and the zero acceleration phases of the swing. Van Eyl concluded that the contribution from the otoliths was negligible, while eye elevation and neck tilt seemed to be the main factors. However, Van Eyl may have failed to find a vestibular effect from the parallel swing because the equivalent head tilt was small (7.5 degrees), and three observers were too few to demonstrate its absence.[18]

Cars have also been used to investigate the effects of linear acceleration in the x axis on perceived size or distance. Goldstein (1959) tested 33 observers in a car which was accelerated forwards or backwards from 0–60 miles per hour in 10–11 seconds (giving an average acceleration of 0.25 g), and then allowed to decelerate more slowly. The visual stimulus was a dimly illuminated red ring, which the observer viewed at 91 cm from his eyes, within a dark box which prevented any peripheral vision. His head was kept still, and he was asked to report any changes in the perceived size or distance of the stimulus. The majority of observers reported that the luminous ring appeared to approach during forwards acceleration and recede during backwards acceleration.

A few reported changes in perceived size instead of, or in addition to, changes in perceived distance. Unfortunately Goldstein did not report the direction of change, or attempt to quantify the effect. We cannot, therefore, say whether the size change was the same as, or opposite to, that reported by Gregory and colleagues (1959) during 'tumbling' in the centrifuge.

Smith (1985) seems to have been the first to perform an experiment that suggests a possible mechanism for an effect of linear acceleration on perceived size. He exposed eight observers to linear acceleration in their x axis. The observers lay on their backs in the dark on a large-stroke vertical vibration platform. They were subjected to sinusoidal oscillation at 0.20–0.67 Hz, with a peak acceleration of 0.10–0.25 g. Records of eye position showed that as the observers were accelerated upwards their eyes tended to converge, and as they were accelerated downwards they tended back towards parallelity. The changes were in the correct direction to compensate for head movements when fixating a static target. This finding, though not directed at the moon illusion, may help to explain any vestibular effect: linear acceleration towards a target can cause convergence, which is accompanied by oculomotor micropsia.

On the whole, the experiments suggest that, during acceleration of the observer, vestibular information may affect size perception. Unfortunately the data from dynamic and static experiments cannot easily be compared. The difficulty lies in trying to obtain an 'instantaneous' measure of perceived size at peak acceleration, and compare it with that obtained during the equivalent static head tilt. It is not at all clear how the difficulties could be overcome.

The view from on high

The question of what is seen out of aircraft and spacecraft is interesting, but hard to interpret, since both the visual scene and the accelerative forces may be unusual. We shall first consider the visual scene and then return to the effects of changes in acceleration.

One might expect the celestial illusion to be more reduced the higher the viewing point above the earth's horizon. It is well known that the perceived size of objects is reduced when observers look downwards, compared with horizontal viewing – an effect that may be related to the 'terrestrial saucer' (Chapter 8), or to the lowered angle of the eyes in the head (Chapter 11), or to empty-field myopia with reduced distance cues (Chapter 7).[19]

Several authors have confirmed that the illusion is reduced from a high viewpoint. According to Schmidt (1964), aircraft pilots reported that some enlargement existed for the moon and stars near the horizon when viewed from a height, but that the illusion increased as the aircraft descended. Schmidt and others also observed a moonrise from a height of 9000 ft (2.7 km), when the moon appeared as a reddish ball above a diffuse layer of haze which obstructed the view of the ground: the moon seemed only slightly enlarged compared to their memory of an overhead moon, but it grew as the aircraft descended. Hamilton (1964) also noticed that the illusion was much reduced when looking down to the horizon moon from aircraft at altitudes of 8000–37 000 ft (2.4–11.3 km). He also carried out a series of experiments (1965, 1966) using an artificial moon to demonstrate that the illusion diminished systematically with height (ground level to 30 m). Hamilton thought that the large illusion at ground level was due to the visible terrain which increased the perceived distance. However, he found that estimates of true distance were either randomly related to viewing height or even increased with height. To preserve the size–distance invariance explanation he suggested that the terrain might cause an increase in *perceived* rather than *estimated* distance. This argument is, of course, similar to the further-larger-nearer hypothesis (Chapter 9).

Ockels[20] was able to observe the setting sun through a dark filter when flying at 39 000 ft (11.9 km), and compare its perceived size with that of a three-quarter full moon viewed without a filter at an elevation of 70 degrees. There was a scattered cloud layer below 20 000 ft (6.1 km), and the horizon was not sharply defined. The sun was at the three o'clock and the moon at the ten o'clock position. As the sun sank from 15 to 2 degrees above the horizon its relative enlargement compared to the moon grew from about 20 per cent to 100 per cent. The perceived size shrank when viewed with one eye only, but a 20 per cent enlargement was still visible when viewed with one eye at 2 degrees. On the other hand, Ross (1982) found that the sun appeared small when viewed over a hazy horizon from an aircraft at 6000 ft (1.8 km), when using both eyes with or without a dark filter. The difference between these observers may lie in the fact that Ockels was viewing the sun almost horizontally over a cloud horizon, while Ross was looking downwards as in the observations of Schmidt and Hamilton.

The views of astronauts

The question is sometimes asked whether astronauts on the moon would experience the celestial illusion: would the earth appear enlarged at earthrise or earthset? Neil Armstrong, the first astronaut to walk on the moon in 1969, was questioned about this: he stated that moon walks took place only during 'daylight' hours, and that crew members were too busy during orbits of the moon to notice the perceived size of the earth at earthrise.[21] John Young, who walked on the moon in 1972, was also questioned: he replied that the moon walks took place in an area near the equator, from where the earth appeared overhead. However, he had been in moon orbit during the Apollo 10 mission in 1969, and had seen many earthrises: he had not noticed any significant enlargement, but this could have been because he was not looking for it.[22] It could also have been because he was looking down towards the moon's horizon (which would reduce any effect), and because the distance to the horizon is shorter on the moon than on the earth (thus reducing the effects of a visible terrain). There is a different report of John Young's statements in another newspaper.[23] He reputedly said that the horizon earth appeared relatively large when viewed by astronauts from the surface of the moon. Other astronauts have reported viewing the raised earth from the moon: 'We could see the earth up there, about the size of a marble', said an unnamed Apollo astronaut in a

NASA film.[24] That report suggests a very small perceived size for the earth seen from the moon – smaller than the estimates of the zenith moon seen from the earth. It is surprising, because the earth subtends a visual angle of about 1.9 degrees (almost four times that of the moon) when viewed from the moon. It would be interesting to know whether most astronauts saw the earth that small, or whether individuals differed in their observations.

A variation on this question is whether the sun or moon would appear enlarged when rising or setting over the earth's horizon, as viewed from a spacecraft orbiting the earth. There appear to be no reports from the early astronauts on this matter. However, Scott Carpenter reported that during his orbital flight in 1962 objects near the horizon appeared relatively enlarged.[25] More recently Byron Lichtenberg reported from the 1983 Spacelab 1 mission that the moon rose so quickly (about 30 degrees in two minutes) that it was not possible to make a size judgement. However, Wubbo Ockels, from the 1985 D1 Spacelab mission, reported that he did observe the moon illusion at moonrise.[26]

Accelerative forces in air and space travel

The effects described in the previous section were viewed under constant gravitational conditions – whether of normal, reduced, or zero gravity. Changes in perceived size have also been observed during aircraft manoeuvres, which involve rapid changes in the accelerative force. Both Schmidt (1964) and Hamilton (1964) noted that the horizon moon, which was much reduced in perceived size when looking down from aircraft, was even further reduced when banking at 60 degrees. Similarly Ockels observed that the moon illusion was reduced by about 20 per cent when he flew at a bank angle of 30 degrees.[27] It is not clear whether the perceived size reduction when banking was due to the tilting of the visual field or to the vestibular stimulation. During banking the increased gravitoinertial force makes the observer feel heavy but upright: the tilt of the aircraft is not perceived, and instead the ground appears tilted (Fig.12.4). The expansion pattern of the ground rising up might induce a compensatory diminution in perceived size.

A contrary observation was made by Galanter, Luce, and Festinger.[28] With the moon at about 45 degrees, they flew a light plane in circles such that at one point the gravity vector was at right angles to the line from the plane to the moon, and at another point the line of sight and the gravity vector were aligned. The observers reported no change in the perceived size of the moon.

More severe changes in acceleration occur during parabolic flight. Parabolas involve a high-g pull-up phase, a zero-g push-over phase, and a high-g pull-out phase. The timing and the g levels can vary, but a typical parabola for a small aircraft is shown in Fig. 12.5.

Ross (1982) made observations as a passenger in a Dornier-128 aircraft flying at about 6000 ft (1.8 km) over Bonn. The aircraft flew parabolas with about 8 seconds of zero gravity, preceded and followed by an acceleration of about 3.5 g. The weather

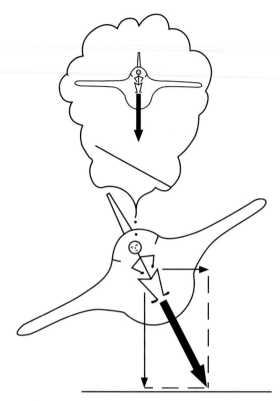

Figure 12.4 Banking and the tilt illusion. During a coordinated turn, the resultant of the acceleration due to the turn and that of gravity is normal to the transverse axis of the aircraft. The direction of this vector is the same as that in straight and level flight, so the pilot perceives the aircraft as horizontal and the ground as tilted. (After Benson 1965.)

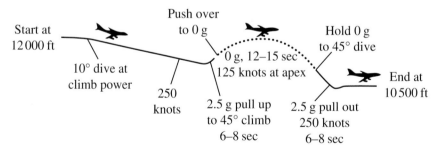

Figure 12.5 A flight profile of a typical parabola for a small aircraft. (After Hammer 1962.)

was clear, but the horizon was hazy. She viewed the high sun binocularly through a dark filter throughout the course of a parabola, but could perceive no obvious changes in size under high or low g. She also obtained a circular afterimage from a flashgun, and viewed the afterimage during a parabola with the eyes shut. Large apparent movement occurred, with the afterimage drifting down with increasing g and up with reducing g.[29] There were no large changes in perceived size, though there was a slight shrinkage at very elevated or depressed apparent locations of the afterimage.[30]

Changes in perceived size seem to occur only during rapid changes in acceleration. The effects are probably due to sensory information that normally indicates bodily movement, and makes the observer compensate for expected changes in the angular size of the target. In orbital flight there is constant zero gravity, and no changes in perceived size have been reported for changes in head tilt.[31] However, it is not known what effect head or body orientation might have on an astronaut's view of the earth. A spacecraft can take various attitudes to the earth and, depending on the location of the windows, an astronaut could adopt various bodily postures to view it. Byron Lichtenberg of *Spacelab* 1 reported that he preferred to look down to the earth rather than up. There is no evidence that zero gravity destroys the moon illusion, as some astronauts report perceiving it when in orbit.

The failure of vestibular experiments

Most of the experiments reported in this chapter seem disappointing. This is not because they demonstrate no effect, but because they demonstrate unclear or unquantified effects. They generally show that tilting the head or body causes an increase in perceived distance and a reduction in perceived angular size. Attempts to distinguish between the effects of otolithic stimulation and those of neck tilt or eye elevation have not met with great success. Stimulation of the otoliths through linear acceleration probably has an effect on perceived size – but in some reports it is hard to determine what the direction of any effect was, let alone its size. Many experimenters were pioneers in using novel and expensive apparatus for perceptual observations, but unfortunately their test procedures and level of analysis did not always match the sophistication of the apparatus. Only very few observers were used, and the authors were usually content to demonstrate the existence of some effect rather than to quantify it. The lack of adequate experimentation can in some instances be excused by the expense of the procedure. It is not easy to obtain research funds to hire an aircraft for observations of the moon, so such experiments are usually performed on an opportunistic basis. Progress in the vestibular and postural field will be expensive, demanding larger numbers of observers and adequate measurements of all the relevant variables.

Despite these strictures, the existing reports imply that further research would be worthwhile. A consistent explanation could emerge, if it could be shown that both linear acceleration towards a target and the equivalent degree of head tilt produce the same reduction in perceived angular size. Such a reduction might occur if both conditions cause convergence of the eyes, thus inducing oculomotor micropsia. Alternatively, micropsia might be directly linked to otolithic stimulation, or to stimulation from the neck receptors, or to any proprioceptive stimulation that indicates that the target is raised or lowered.

Visual and tactile space

In this book we have mainly been concerned with the visual judgement of size, since the moon illusion is a visual illusion. We have said little about tactile judgements of size, since celestial objects are beyond the reach of the hand.[32] Yet for close objects vision and touch interact: visual size judgements guide the hand to grasp an object correctly; and tactile size judgements serve to correct vision when it errs.

Many authors believed that the touch sense was less fallible than vision, and that touch educates vision. Plotinus wrote in the third century: 'Touch conveys a direct impression of a visible object.'[33] The question of whether tactile and visual size were innately linked was discussed by Molyneux in the seventeenth century and became known as *Molyneux's question*;[34] many inconclusive investigations were made of formerly blind people who recovered their sight. Berkeley continued the debate and argued that tactile experience enables us to interpret the visual image size and perceive the true linear size, even for objects beyond the reach of touch. He argued that both the visible size (image size) and the tangible size (physical object size, whether tactile or visual) were open to perception, but especially the latter as it was of more practical importance: 'Hence it is, that when we look at an object, the tangible figure and extension thereof are principally attended to; whilst there is small heed taken of the visible figure and magnitude, which, though more immediately perceived, do less concern us, and are not fitted to produce any alteration in our bodies' (1709, Section LIX). The horizon enlargement of the moon, he argued, was like the enlargement of a real tactile object rather a visual image: 'When therefore the horizontal moon is said to appear greater than the meridional moon, this must be understood not of a greater visible extension, but of a greater tangible or real extension' (1709, Section LXXIV). What exactly Berkeley meant by this remains obscure.

In the twentieth century there was much interest in the perceptual effects of optical distortion. Experimental studies disproved the inherent dominance of touch, and showed instead the phenomenon of *visual capture*: the hand feels to be the size it looks. However, after some time, both vision and touch are modified, and a new compromise between the senses is reached.[35] It is usually argued that the discrepancy between the senses is what drives the adaptation process. However, adaptation occurs even though the observer is not consciously aware of the discrepancy – which suggests that different size values can be held at different levels of consciousness in the visual and tactile perceptual systems. Indeed, other experiments with undistorted vision show that visual and tactile size are not always identical.[36] On the other hand, many researchers use unseen tactile adjustments as a way of measuring visual perception – which suggests that they believe that the two systems share the same size values.

It has recently been shown that the brain has two visual pathways leading from the primary visual cortex: a ventral stream serving perception and cognition and a dorsal stream serving action.[37] Size perception and object identification are usually thought to belong to ventral stream areas, while the localization of targets in space, and the guidance of movement, are relegated to the dorsal stream (Fig. 12.6). At close distances

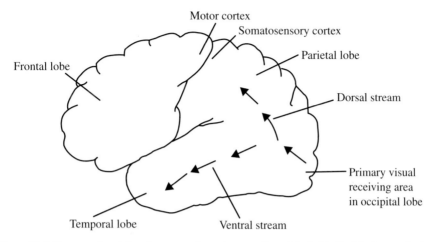

Frontal lobe

Motor cortex

Somatosensory cortex

Parietal lobe

Dorsal stream

Primary visual
receiving area
in occipital lobe

Temporal lobe   Ventral stream

Figure 12.6 The ventral and dorsal streams, shown in an outline of the left half of the human brain.

both the dorsal[38] and the ventral[39] streams are normally geared to linear size rather than to angular size.

Usually the ventral and dorsal streams compute size fairly consistently, but discrepancies may sometimes arise. We have already mentioned the example of optical magnification. It has also been claimed that distortions of a purely visual origin can have different effects on vision and touch. For example, size-contrast illusions may affect visually perceived size but not manually adjusted grip size.[40] However, other authors dispute this;[41] and recent reviews[42] suggest that the effects vary with the method of measurement and the nature of the task. It would indeed be strange if the primary visual cortex always fed accurate information to the dorsal stream but misleading information to the ventral stream. Even if this is the case, it does not prove that perceptually conflicting size values can be held *within* the ventral steam. The size of the moon remains beyond our grasp.

Another controversial question is whether different perceived distance values can be held in the two streams. One aspect of this is whether visually perceived size can be scaled to distance as perceived by the tactile–kinaesthetic system: there is some conflicting evidence concerning afterimages viewed in the dark and imagined to be projected onto the hand,[43] and also concerning judged size in computer displays.[44] A more direct comparison of visual and tactile distances was reported by Mon-Williams and Tresilian (1999), who found that convergence of the eyes produced verbal reports of micropsia and of increased distance, whereas hidden reaching to the same perceived distance produced a 'near' response (Chapter 9). They argued that the verbal reports were cognitive judgements based on the perceived size change, while the reaching responses were cognitively impenetrable. To make some sense of this in terms of two visual streams, we should have to say that the oculomotor system sends accurate distance information to the dorsal stream, but sends inaccurate information about angular size (and consequently distance) to the ventral stream. However, the moon is so distant that it must be perceived by the ventral stream alone – and it remains uncertain whether conflicting perceived distances can exist within that stream.[45]

Summary

Knowledge of vestibular and other proprioceptive mechanisms lagged behind that of vision, arriving in the late nineteenth century. Experiments involving head and body tilt show that tilting backwards can enlarge perceived distance by up to 27 per cent. Perceived angular size is probably reduced by a much smaller amount. Size matching experiments across orientations suggest a reduction of up to 22 per cent, but the lack of counterbalanced designs makes this value uncertain. The size effects may be caused by oculomotor micropsia, or may be conditioned to any stimuli indicating gaze orientation with respect to earth's gravity. Experiments involving vestibular stimulation through applied accelerations in various forms of travel have not provided clear effects on perceived size. Observations made from aircraft and spacecraft are also inconclusive. Experimental methods in all these areas have been inadequate. The relation between tactile–kinaesthetic space and visual space is controversial, but may not be relevant to the celestial bodies which can only be perceived in visual space.

CHAPTER 13

Conclusions and conundrums

We look back on the history of the moon illusion with mixed feelings: a slight sense of disappointment, some pride in what has been achieved, and considerable hope for the future. Disappointment arises from the seemingly endless repetition of theories, and the slow growth of factual knowledge. The story starts with a few lone scholars, from whose works we have quoted, often at length; but by the twentieth century we are faced with a veritable army of experimental psychologists, whom we scarcely have room to mention. Despite all the effort, much confusion remains. Psychologists still have no agreed terminology for matters concerning perceived size, and they continue to argue in slippery language over disputable facts. They cannot agree whether or how distance cues play a role in determining perceived size, or whether this role is the same for near and far distances. We envy biochemists who can unravel the structure of DNA, and astronomers who can calculate the age of the universe.

This may be too bleak a picture. The physical sciences are not as solid as bedrock, and the study of human perception is not a hopeless enterprise. Kitty Ferguson (1999, pp. 297–8), in her book on the history of the measurement of the universe, points out that scientists have hesitatingly groped their way towards their current ideas, and that what they have discovered so far is only a small island of knowledge in a sea of uncertainty. She goes on to say that we human beings are 'the most complex thing we have yet discovered in the universe. The human mind is still largely unexplained.' So perhaps the trouble with visual perception is just that it is a more difficult enterprise than astronomy. Progress on the moon illusion and size perception may be slow, but many other aspects of visual perception (such as stereopsis, colour vision, and movement perception) are now well understood at a neurophysiological level. Advances in perceptual knowledge have come from the development of better apparatus, the use of psychophysical measuring techniques, the employment of large numbers of observers and the application of statistical significance tests; and finally by the linking of established perceptual phenomena to neurophysiological mechanisms. In the case of the moon illusion, the replacement of anecdotal observations by careful measurements has made it possible to discard many explanations and to refine others. We now summarize our conclusions.

Explanations that must be rejected

In the course of this book we have reviewed a large number of theories about the moon illusion, and described a great many observations and experiments. No single theory has emerged victorious, but some explanations have been ruled out. We argued in Chapter 3 that theories could be divided into three classes: (1) image size changes for external physical reasons; (2) image size changes within the eye for reasons connected with physiological optics; (3) perceptual size changes owing to scaling mechanisms within the brain.

We can eliminate the first two classes without much difficulty. It is easy to dismiss the idea that there is some change in image size owing to physical causes, such as a change of real distance (Chapter 2) or atmospheric refraction (Chapter 5). Astronomical knowledge rules out the first, and physics the second. If there is any doubt in one's mind, photographs of the moon show no discriminable change in image size with elevation. It is harder to show that there is no change in image size owing to processes within the eye, because the evidence is less accessible; but it is unlikely that any major changes occur, or could contribute to the horizon enlargement (Chapter 7). One such explanation is based on the expansion of pupil size under low illumination, which might cause inadequate focusing and a blurred image. The luminance of all celestial bodies is indeed lower on the horizon than higher in the sky, but the sun remains so bright that it is always viewed with constricted pupils. Other theories based on faulty accommodation are unconvincing, particularly when applied to the sun and constellation illusions. None of these theories can account for the afterimage illusion, because afterimages are unaffected by the optics of the eye. A decrease in the luminance of the moon and sun near the horizon may cause a small reduction in image size, owing to the reduction of irradiation within the eye (Chapter 6), but such an effect is opposite to the direction of the illusion. Theories involving binocular disparities are unlikely to be relevant to very distant objects (Chapters 6 and 11), and cannot explain the monocular illusion. In summary, it seems unlikely that image changes within the eye can provide even a partial explanation of all the celestial illusions.

Factors relevant to the moon illusion

There remain several perceptual factors – and several explanations – that could apply to the moon illusion. These need not be mutually exclusive. Indeed there is a long tradition, dating back to classical times, of listing several contributing factors. Careful measurement in the twentieth century has shown that some of the traditional factors do affect the moon illusion, though other factors still await confirmation. It is useful to have a quantitative description of the possible factors, to show that the combined effect is sufficient to produce the illusion. The relevant factors fall into three main groups: (1) relative size and terrain effects (Chapters 9 and 10); (2) vergence commands, angle of regard, and posture (Chapters 7, 11, and 12); and (3) aerial perspective and colour (Chapter 6).

Relative size and terrain effects

The scaling effect of the terrain is the best-established contributing factor to date. The presence of terrain has been shown, in experiments with artificial moons, to cause perceived enlargements by amounts varying from 26 to 66 per cent. The effects of increased terrain and intervening objects are large enough to be quite noticeable, and they form the basis of many observational reports. The celestial bodies are perceptually enlarged near the horizon when a large extent of terrain is visible, probably because they are scaled to the same extent as other objects in that part of the visual scene. The size enlargement does not depend purely on local contrast with objects of small angular size, because the horizon moon is seen as enlarged across an empty expanse of water, and the raised moon as small even when surrounded by small stars. The scale seems to be determined by relative size over the whole visual scene, particularly the foreground terrain. A small moon illusion can be demonstrated in pictorial representations, similar to geometrical illusions of size; but the pictorial effect is not purely geometrical, because inversion of the display reduces the perceived size difference. The real moon illusion is also reduced when the observer inverts his head, and the artificial illusion is reduced when the image of the real terrain is inverted by means of prisms. The terrain effect may therefore be affected by height in the visual scene, or by some cognitive interpretation of the scene. Many authors argue that the sight of the terrain increases perceived size because it increases perceived distance – an argument we shall revisit.

Vergence commands, angle of regard, and posture

Factors connected with eye control and bodily posture may contribute a little to the celestial illusion. Well-designed experiments with large numbers of observers show that raising or lowering the eyes in the head reduces the perceived size of artificial moons by about 2–7 per cent, and also makes them appear further away in comparison with horizontal viewing (Chapter 11). Tilting the head or body has similar results, though the size of the effects has not been rigorously measured (Chapter 12). Unlike the sight of the terrain, these angle of regard effects are so small that they are not readily noticeable by the casual viewer – instead they have to be deduced from large experiments. A noticeable size reduction occurs only when the observer inverts his head by looking through his legs or lying on his back; and in that case the effect is probably caused by inversion of the visual scene on the retina.

   Explanations of the small eye level and postural effects are uncertain. The changes in perceived size are not caused by any measurable changes in image size; they are probably conditioned to eye, head, and body posture, or to the oculomotor commands that accompany postural changes. The purpose may be to reduce the perceived angular size of objects overhead or underfoot (which are usually near and of large angular size) in order to assist size constancy. Several oculomotor factors may be involved, because vergence, accommodation, and pupil size change together in a reflex manner: the 'near' response (convergence) is accompanied by a diminution in perceived size, while the opposite happens for the 'far' response (divergence). It is interesting to note that the perceived distance effects of the oculomotor changes are also paradoxical, in that the judged distance is usually the opposite of the actual vergence response; this may give some empirical support to the idea that two conflicting types of distance judgement occur at different levels of consciousness.

Aerial perspective and colour

There is plenty of anecdotal evidence linking a light haze and a reddish colour to an extra large appearance of the sun or moon, but there is currently no experimental evidence to show that either factor contributes to the celestial illusion. A thick mist can indeed cause non-luminous terrestrial objects to be judged larger in linear size by 20 per cent or more, but the same may not apply to luminous objects. The sun and moon actually appear smaller in such conditions, because their luminance is reduced, and dimmer objects appear smaller. However, experimental measurements are needed to compare the size of the moon illusion in the same terrain in both clear conditions and a light haze. Similarly, the colour red requires further investigation. Some laboratory experiments suggest that a red colour may increase perceived size by about 3–6 per cent, though it has not been shown to do so in a moon machine experiment. Experiments are needed in which a simulated moon is coloured red, but matched for perceived brightness with a white moon before size judgements are made. A thorough series of experiments on colour, brightness, and haze would reveal whether atmospheric variation contributes to the variability of the illusion on different occasions, or whether the reported link is coincidental.

The magnitude of the celestial illusion is an elastic quantity, because it varies with the viewing conditions and with the manner of measurement. A typical average value is a horizon–zenith ratio of about 1.5, or an increase of 50 per cent. Individual observers sometimes give much higher values, but averages for large groups do not usually exceed a ratio of 2.0, or an increase of 100 per cent. The magnitude of the contributing factors is, of course, as elastic as the illusion itself: these factors are subject to many of the same sources of variability as the celestial illusion. However, the largest contributing factor seems to be the sight of an extensive terrain, which can cause enlargement of 40 per cent or more; it is thus sufficient to account for most of the celestial illusion. In addition, oculomotor commands and eye/head/body position probably contribute up to 10 per cent. The effects of a light haze and a red colour await satisfactory measurement, but anecdotal reports suggest that the effect is greater than 10 per cent. A total effect of 60 per cent or more is quite sufficient to account for the normal celestial illusion. If the factors combine in a multiplicative manner, they could produce even larger illusions.

Perceptual explanations

The factors listed above all depend on perceptual processes, but their explanation remains controversial. Many authors have argued that the contributing factors affect perceived distance, which in turn affects perceived size. The argument has been applied to intervening objects, the apparently flattened sky, aerial perspective, postural factors, and oculomotor effects. The theories claim that one or more of these factors make the celestial bodies appear further away on the horizon, and therefore larger. The explanation runs into difficulties, at least when combined with the classical version of size–distance invariance. The role of the flattened sky dome is dubious, because the sky does not always appear flattened, and because the celestial bodies appear to float behind or in front of the dome. Moreover, the small estimated linear sizes of the sun and moon on the horizon are not geometrically compatible with a large perceived distance (Chapters 3 and 10). However, the main reason why perceived distance theories are unacceptable is that most people report the celestial bodies as appearing closer and not further away when rising or setting.

This last fact is a very awkward one for many perceptual phenomena, and is known as the size–distance paradox. Theorists attempt various ways around the difficulty; they propose two levels of processing (automatic and conscious), or two types of size perception (angular and linear), or two types of distance perception (registered and judged); and they mix and match these in various combinations. We have frequently mentioned the two-sizes approach (Chapters 3 and 4), which developed during the twentieth century. The two-distances approach has earlier origins, and seems to depend on the idea of an unconscious inference, which is often attributed to Locke or Helmholtz (Chapter 9). We have labelled it the *further-larger-nearer* theory, to draw attention to its paradoxical nature. Proponents of this theory explicitly state that two distance judgements of the moon are made at different levels of consciousness or reasoning; but they tend to be inexplicit about whether two types of size perception are involved. The first distance judgement, variously described as subconscious or as conscious, is that the horizon moon appears further than when elevated, and this causes an increase in its consciously perceived (linear) size. The second distance judgement, made consciously, is that the horizon moon appears closer, because the perceived (angular) size of the horizon moon is enlarged. The implicit confounding of angular and linear size causes difficulties for classical size–distance invariance. Many theorists also object to the idea that observers can 'perceive' or 'register' the horizon moon as far away at one level of consciousness, but 'perceive' or 'judge' it as near at another level of consciousness. In the absence of empirical evidence, the competing theories seem like redescriptions of the size–distance paradox. Nevertheless, recent experiments by Kaufman and Kaufman (2000) provide evidence that the binocular stereoscopic system treats the artificial horizon moon as further away than the elevated moon (Chapter 9); and it is possible that the same may apply to natural viewing of the real moon. Moreover, current work on visual illusions suggests that different size values and different distance values can be held in the ventral and dorsal pathways in the brain; and while this work is not directly relevant to the moon illusion, it does give some credence to the idea that different spatial values can be held at different levels of consciousness (Chapter 12).

The further-larger-nearer theory is in part a cognitive theory, since it implies that the second distance judgement depends on 'conscious' reasoning. However, this conscious reasoning does not operate in a precise geometrical manner, since the small estimated linear sizes of the celestial bodies are incompatible with their large estimated distances. Another cognitive theory holds that the celestial bodies are compared to large objects of known size on the horizon (Chapter 10) – but again there is no logical relation between these size estimates, and substantial illusions occur in the absence of such objects. Cognitive explanations might be given some support by the occasional reports of huge illusions, when the sun or moon is assumed to be some other object (Chapter 1); but in these instances the cognitive errors are so implausible that it is more likely that the large illusion caused the cognitive error than vice versa.

The moon illusion, angular enlargement, and size constancy

Many authors have argued that the moon illusion is an example of size constancy – but that begs the question as to the nature of size constancy. 'We might consider the story of the efforts to understand the moon illusion as a microcosm of the larger problem of size constancy itself', wrote Dember and Warm (1979, p. 209). Those authors who believe that size constancy is achieved by 'taking account of distance' usually give the same explanation for the moon illusion; and those who believe that size constancy is achieved in other ways usually give other accounts of the moon illusion. Most investigators agree that the horizon moon appears enlarged by a factor of 1.5–2.0 compared to the raised moon, but they disagree as to whether this enlargement is primarily linear or primarily angular, or whether such a distinction even exists.

If the horizon enlargement is regarded as angular, it should be noted that its value is less than half of the enlargement of 4.0 or more found for the perceived angular size of terrestrial objects at far distances (Chapters 3 and 4). The discrepancy probably occurs because of the different baseline values. The appearance of the raised moon is normally taken as the baseline when measuring the moon illusion, but the raised moon itself is usually judged as larger than a near target of the same angular size (Chapter 4). The appearance of a near adjustable target is taken as the baseline in many angular size experiments, and the distance of that target clearly affects the resulting ratio. If the perceived angular enlargement of the horizon moon is measured in the same way as that of other horizon objects, the enlargement is found to be about the same as for those objects.

We have already rejected the idea that the horizon moon's perceived enlargement should be described solely as a linear enlargement, being caused only by an increase in consciously perceived distance. A different interpretation is that the enlargement is indeed linear, but represents some combination of perceived angular size and perceived distance. According to McCready (1986) the horizon moon typically appears to be at half the distance of the zenith moon and linearly enlarged by a factor of 1.5, which implies an angular enlargement by a factor of 3.0. This factor is closer to that noted above for experiments showing the growth of angular size judgements with distance – but only if the latter data are treated as pure angular enlargements and not as some combination of angular and distance effects. With a shortage of pure angular measures, we are left with some uncertainty as to the extent of the angular enlargement both for the moon illusion and for other size–distance experiments.

Provided the same method of measurement is used, the size of the angular enlargement is broadly similar for the horizon moon and for terrestrial objects near the horizon. This finding has two implications. First, terrestrial size constancy – like the moon
illusion – can be regarded as consisting mainly of an enlargement of perceived angular
size. Within close distances this effect is so strong that objects do not appear to grow
much smaller with distance; but at far distances objects do appear smaller, and
judgements of true linear size must be based on cognitive calculations rather than
appearances. Second, much the same factors that contribute to the moon illusion
also contribute to size constancy for distant objects. The factors are probably not
identical, but there is considerable overlap. Differences in the contributing factors may
account for any differences in the size of the enlargement that cannot be explained by
measurement artefacts.

Remaining tasks and problems

Perceptual psychologists who work at the classificatory level of explanation still have
much to do in the way of definition and measurement. It would be desirable to set up
an authoritative committee to define the acceptable terms, formulae, and measurement
procedures for spatial perception. Another step would be to persuade a large number of
laboratories to embark on an agreed series of experiments to measure the effects of
various factors on perceived linear size, angular size, and distance. Such a project would
be much cheaper than the Human Genome Project, and the results would remove much
of the uncertainty about spatial perception. Unfortunately, because perceptual research
is relatively cheap, individuals can continue their investigations in an anarchic manner.
While 'Blue Sky' research can be brilliant, it is unlikely to clarify the moon illusion.

The concept most in need of further definition and measurement is 'perceived angular size'. The term is often used as something of a rag-bag to cover any type of perceived
size that does not obey classical size–distance invariance. Most investigators of perceived
size have used linear matching techniques while combining them with either angular or
retinal size instructions (Chapter 3). To test whether angular and retinal size can be
regarded as perceptually equivalent, experiments should also be conducted using truly
angular techniques. These techniques include numerical estimation in degrees, the
matching of sizes to angles on a protractor or other device, and the production of
sketches showing the perceived proportions of the visual field at different distances.
Pointing techniques can also be used for very large angles, but not for angles as small as
that of the moon. Currently there is inadequate experimental evidence, and insufficient
theoretical clarity, to determine whether the moon illusion should be described as one
of perceived angular size, or perceived retinal size – or perhaps neither.

There is still a need for more experiments that use large numbers of subjects and
adequate designs. Some of the experiments could be technically simple, but others need
complicated apparatus. Many of the early moon illusion experiments should be
repeated, and new ones should be devised; but measurements should be obtained for
more hypothetical types of perceptual size (linear, angular, and retinal) and distance

(registered and judged). It is also important that adequate experiments should be run at far distances, rather than confined to a laboratory. The appearance of distant natural objects can be measured by numerical estimates.[1] Apparatus recently developed by Kaufman and Kaufman (2000) has made it possible to investigate the registered distance of an artificial moon, projected stereoscopically to a distant horizon or higher in the sky, by adjusting the stereoscopically perceived distance of a second artificial moon to bisect the distance of the far moon (Chapter 9). Additional methods need to be devised for measuring both the registered and the judged distance of the natural moon. Modern equipment should also make it possible to run outdoor-size matching experiments in which the distant target is adjustable in addition to the near target, thus controlling for the tendency to set the variable target too large (the 'error of the standard' – Chapters 3 and 4). Better-designed experiments should reveal whether or not the various types of judgement correspond to geometrical principles – but we suspect the outcome would be negative.

Abandoning geometry

The application of geometry was one of the success stories of science. Geometry enabled early astronomers to measure the earth, sun, and moon. It formed the basis of perspective drawings, and it later enabled scientists to understand the optics of the eye and to build telescopes. It is not surprising that it was adopted as *the* model for the perception of size and distance. The doctrine of size–distance invariance has ruled our thoughts from at least the time of Posidonius to the present day (Chapter 3). One of the few challengers was Berkeley, who inveighed against the geometrical theorists whom he called 'the optic writers'. But such is the sway of geometrical optics that Berkeley's opinions are today often quoted (by those who have not read them) as though he upheld size–distance invariance.[2] Even the recent challenge by McCready and others, with the concept of perceived angular size, is yet another form of size–distance invariance. We have to consider whether any form of size–distance invariance makes a useful contribution to the understanding of visual space perception.

All versions of size–distance invariance assume that both angular and linear size can be perceived at some level of consciousness or at some time. The important difference between the classical and perceptual version is that angular size is usually taken to be correctly perceived in the former but subject to error in the latter. Nevertheless, many people find it difficult to accept that two different size values, or two quite different sorts of size, can be perceived simultaneously. If angular size is like retinal image size, one would have to say that a small linear image (equivalent to a near projection of the retinal image) is perceived at the same time as a large linear image (equivalent to a far projection of that image). If angular size is like a subtended angle, one would have to say that some angular aspect of the scene is perceived in addition to linear size. These views are hard to understand, and there is little introspective or experimental evidence to support either of them.

One response to these difficulties is to reject all geometrical accounts, and to say that visual size and distance are computed independently of each other.[3] The same factors are normally used in the computation of size and distance, giving a correlation that approximates to classical size–distance invariance; but in some situations different factors are used for size and distance, giving rise to size–distance paradoxes such as the moon illusion. This approach has the advantage of abolishing the need for more than one type of visually perceived size or distance.

If there is only one type of perceived size, we have to ask how the different judgements of angular and linear size are obtained. Is angular size the primary percept and linear size a secondary calculation, or is it the other way round? Or is some unspecified perceived size the primary percept, with both angular and linear size being secondary calculations? Recent experiments at short viewing distances have given contradictory results,[4] and do not help to resolve the issue for far distances as in the moon illusion. It will be a challenge to devise comparable experiments at far viewing distances.

In the previous chapter we discussed the relation between two other very different sorts of size – visual and tactile. There is some evidence that different size values can be held simultaneously in these two systems, which might give credence to the idea that the same can happen *within* the visual system. Unfortunately the tactile evidence relates only to close viewing distances, and cannot help to elucidate far distance perception. Perhaps some ingenious investigator will find a way of applying tactile space to far distances.

The future of the moon illusion

The moon illusion is one of the few perceptual phenomena that tap a broad spectrum of sciences: astronomy, optics, physics, physiology, psychology, and philosophy. Its explanation illustrates the history of scientific explanation, and in particular the history of perceptual psychology. It fails to follow the latter in detail because the illusion keeps being rediscovered or re-explained by 'laymen'. These authors start afresh with a small number of ideas that have been in circulation for centuries. They have no difficulty in re-inventing size constancy or in suggesting that atmospheric refraction is the cause. Indeed, popular newspaper columns continue to promulgate the latter idea.[5] However, the growth of psychology courses in universities has had the effect that some knowledge of perceptual theories has infiltrated the general literature. We give one example in the epilogue of a poem on the moon illusion by Helen Bevington which mentions the effects of bodily posture. We give here an excerpt from a novel by Kate Atkinson (1997) which mentions both bodily posture and visual surrounds. Some friends have noticed a huge horizon moon, and discuss its cause:

> Eunice pipes up, 'We're only experiencing the moon illusion – it's an illustration of the way the brain is capable of misinterpreting the phenomenal world.'
>
> '*What?*'
>
> 'The moon illusion,' she repeats patiently. 'It's because you've got all these points of reference' – she waves her arms around like a mad scientist, 'aerials, chimney pots, rooftops, trees – they give us the wrong ideas of size and proportion. Look,' she says and turns round and suddenly bends over like a rag doll, 'look at it between your legs.'

'See!' Eunice says triumphantly when we finally obey her ridiculous command. 'It doesn't look big any more, does it?' No, we agree sadly, it doesn't.

'You lost those points of reference, you see,' she carries on pedantically.

Laymen can easily explore the earlier theories: anyone can manipulate mirrors, view the moon with different bodily orientations, project afterimages onto the sky or observe the apparent shape of the heavenly vault. It is only in recent years that the apparatus has become more technical, the experimental design and statistics have become more complicated, and the level of theorizing has gone beyond the reach of the average layman. The typical layman does not appreciate the mathematics of adaptation-level theory, Fourier analysis, or accelerative forces, nor the intricacies of the vestibular system. He does not have access to a centrifuge or parallel swing, and cannot repeat the experiments described in Chapter 12. Visual research in particular has become so scientifically technical that much of it is beyond the comprehension of the average psychology student with an arts background.[6] Though there is still a need for technically simple experiments, it is unlikely that laymen will be able to design and conduct them.

The illusion continues to be a topic of interest to philosophers of perception. Until recently, there has been little interplay between the philosophical and psychological literature. For example, Heelan (1983) wrote a book on *Space-perception and the philosophy of science*, in which he applied his ideas on hyperbolic visual space to the moon illusion and other natural outdoor phenomena: however, he described little experimental evidence and relied heavily on the popular writings of a few psychologists. Psychologists are equally guilty of ignoring the work of philosophers. None of the contributors to Hershenson's (1989) book on *The moon illusion* mentioned Heelan's theory, and we were alerted to it by the work of a Dutch classical scholar, Loek Schönbeck (1998). Psychologists tend to feel that they have nothing to learn from philosophers; but several, such as Schönbeck, point out the importance of a clearly defined terminology for perceptual matters. Recent philosophers such as Schwartz (1994) show detailed knowledge of the psychological literature, and make insightful comments on perceptual theories. The growth of the internet has meant that academics from different disciplines can now easily discover each other's work.

There are grounds for optimism over future studies of size perception. Hershenson's (1989) book contains a wide variety of interpretations, and most authors accept that the moon illusion is not explained in a simple manner by perceived distance. The perceived angular size account is now included in some textbooks,[7] and there is also a great deal of research activity on conflicting spatial judgements in different visual pathways in the brain. A recent book on the constancies has been edited by Walsh and Kulikowski (1998), and it contains some exciting new approaches to the neural encoding of size. Developments in neuroscience are now very rapid, and we should soon know more about the parts of the brain involved in size and distance perception, and the neurological mechanisms by which the representation of the retinal image is rescaled along the route to conscious perception. We tend to use the term 'illusion' for phenomena that we do not fully understand; so once the neurological mechanisms are understood, the celestial illusion – and a host of other size illusions – will cease to be called illusions. The sun will have set on the moon illusion.

Summary

The measured size of the moon illusion typically shows a horizon enlargement of about 50 per cent compared to the raised moon, but sometimes much more. It can be accounted for as the sum of several factors. The most important of these is the visible terrain, or relative size effects over the whole visual scene, which can cause size enlargement of about 40 per cent. Oculomotor commands, angle of regard, and posture probably contribute another 10 per cent. The effects of a light haze and a red colour have not been adequately measured, but may contribute another 10 per cent or more. These factors serve to increase the perceived size of the horizon moon relative to the raised moon, whilst also making it appear closer. This relationship contravenes classical size–distance invariance. Perceptual size–distance invariance can be preserved if the illusion is described as involving a change in *perceived angular size*. However, the meaning of this term is obscure, and its measurement elusive. It remains uncertain whether both angular and linear size can be visually perceived. It may be necessary to abandon the geometrical approach to spatial perception, and to treat size and distance as independent entities that can be processed at more than one level of consciousness. Recent evidence gives some support to the idea of conflicting spatial representations in different visual pathways. When measured in the same way, the perceived enlargement of the low moon is found to be the same as that of terrestrial horizon objects; therefore, whatever analysis is given to the moon illusion, the same should be applied to normal size constancy. There is a need for further well-controlled experiments on different types of perceived distance and perceived size; and these should be conducted at both near and far distances, using several different methods. The next few years are likely to see large advances in the neuroscience and psychophysics of size perception.

Epilogue

The moon illusion remains a popular topic of conversation among the general public. Whilst the erroneous idea of atmospheric refraction remains entrenched, versions of the angle of regard or bodily posture theories are gaining ground. We quote the following poem by Helen Bevington.

Academic Moon

I have been walking under the sky in the moonlight
With a professor. And am pleased to say
The moon was luminous and high and profitable.
Moonlit was the professor. Clear as day.

He had read, of late, how extraordinary moons are
Upside down. Aloft in the night sky
One drifted upright, in the usual fashion.
But the professor, glad to verify

Hypothesis or truth, when he is able –
Even, it seems, to set the moon askew –
Proposed that we reverse our own perspective.
And, on the whole, it was a lovelier view

Of white circumference – smaller now, he fancied,
A tidier sphere. This last I could not tell
From so oblique an angle. I only remember
Enjoying the occasion very well.

Appendix: Summary of scientific developments relating to the moon illusion

Prehistoric period

3000–2000 BC Late Neolithic people built stone circles, such as Stonehenge in England, which acted as calendars and show systematic interest in the movements of the sun and moon.

Classical period (700 BC–AD 200)

The scientific tradition

600–300 BC

A scientific world view developed in Greece: for the first time philosophers attempted to give a natural, rather than a supernatural, explanation of the world. The climax of this Golden Age was marked by the works of Aristotle (350 BC).

300 BC–AD 200

In spite of the development of natural philosophy, science and magic were for many centuries not seen as separate provinces of knowledge. This is evident in, for example, Pliny's 37-volume *Natural history* (AD 77).

AD 0–200

Greek scientific knowledge reaches China.

The solar system

700 BC

Arithmetical methods to describe the irregular motions of the sun, moon, and planets were developed by the Babylonians for astrological purposes. The systems of time and angle measurement used by them (based on much older Babylonian systems of measurement) have remained basically the same to the present day.

550 BC

Anaximander pictured the heavens as a sphere rotating around the earth on a fixed axis.

500 BC

Pythagoras taught that the earth was spherical, that the morning and evening star were the same planet, that the moon's orbit was not in the plane of the earth's equator, and that the sun, moon, and planets all travel in their own separate paths. He proposed separate celestial spheres to carry them – a notion that survived to the sixteenth century.

450 BC

Anaxagoras explained the phases of the moon and eclipses.

350 BC

Heracleides suggested that the apparent daily rotation of the heavens could be explained by the rotation of the earth on its axis, and that Mercury and Venus revolve around the sun. These views were not generally accepted until the time of Copernicus in the fifteenth century.

270 BC

Aristarchus suggested that all the planets revolve around the sun, but his theory, like that of Heracleides, found no favour. He also attempted the first scientific estimate of the relative distances to the moon and sun.

240 BC

Eratosthenes made the first scientific (and reasonably accurate) estimate of the size of the earth.

130 BC

Hipparchus developed a model of planetary motion in terms of deferents and epicycles; compiled the first star catalogue, containing the positions and brightnesses of 1000 stars; and made the first reasonably accurate measurement of the moon's parallax (and therefore its distance from the earth). He knew that the moon's distance was not the same at all points in its orbit.

AD 140

Ptolemy's *Almagest* was published: it remained the standard work in astronomy to the fifteenth century. Ptolemy rejected the rotation of the earth and the theory that some or all the planets move around the sun. As a result these theories were not generally accepted for the next 14 and 15 centuries respectively.

Optics, physiology and vision

450 BC

Alcmaeon discovered the optic nerve, and concluded that the brain is the central organ in sensation and perception. Empedocles' theory of vision: visual rays emitted by the eye fall on the object and return to the eye to cause a visual image. This theory remained important throughout classical times.

430–330 BC

The Hippocratic medical corpus was written by various authors, establishing empirical medicine and medical ethics.

320 BC

Euclid's book on optics claimed that perceived size is equivalent to visual angle.

300 BC

Hirophilus dissected the brain and nervous system, distinguishing between sensory and motor nerves.

100 BC

The refraction of light was recognized.

AD 140–170

Ptolemy studied the refraction of light (in water), and discussed many perceptual effects including the scaling of image size for perceived distance.

c. 200

Cleomedes argued that perceived size is equivalent to linear size scaled for distance.

150–200

The Greek physician Galen studied the brain and nervous system through dissections. His description of the anatomy of the eye was accepted for several centuries, until improved upon by Arab scholars.

Arabic science (750–1200)

750–800

Origin of Arabic science.

813–833

During the reign of Caliph al-Mamun many classical Greek works were translated into Arabic in Baghdad, including Ptolemy's *Almagest* (by the Jew Sahl al-Tabari) and the works of Plato and Aristotle. Hunayn ibn Ishaq translated the medical texts of Galen and the Hippocratic corpus.

c. 830

Al-Kwarizmi compiled his book on algebra, in which Hindu numerals, including zero (now usually referred to as 'arabic numbers') were introduced to the Arab world.

800–1050

Arabic science surpassed that of any other culture. The climax of Arabic learning was reached during the first half of the eleventh century.

1040

Ibn al-Haytham, perhaps the greatest physicist of medieval times, investigated the rainbow, estimated the height of the earth's atmosphere and improved understanding of the anatomy of the eye. Like the ancients, he regarded the lens as the sensitive organ of vision. He rejected the visual ray theory in favour of the idea that vision occurs when light scattered from objects enters the eye. He also discussed many perceptual effects, including the scaling of image size for perceived distance.

Science in medieval Europe (900–1500)

The scientific tradition

976

Gerbert of Aurillac (later Pope Sylvester II) introduces Hindu–Arabic numerals (except for 0) to the diocesan school of Reims, but with little effect. Subsequent translations into Latin of al-Kwarizmi's book on algebra (e.g. by Robert of Chester, *c*. 1100) and books on arithmetic with the new numbers (e.g. by Leonardo Fibonacci of Pisa, 1202) led to their gradual adoption in Europe over the next few centuries, making arithmetic much easier and stimulating progress in mathematics.

1100–1200

Many important classical and Arabic works were translated from Arabic into Latin for the first time, especially in Toledo, by Gerard of Cremona (1114–87) and others: these works included Ptolemy's *Almagest* and works of Aristotle, Galen, and the Hippocratic corpus. During this and the next century there was much interaction between the declining Arabic science, Jewish scholarship, and emerging Christian science. This led to the supremacy of Latin as the scientific language for three centuries, and the scientific supremacy of western Europe.

1200–1300

Rise of the universities in Europe, continuing in the next century.

1300–1400

Direct translation of classical works from Greek into Latin commenced, while translation from Arabic declined.

1450

Printing with moveable type was introduced in Europe. It soon spread throughout Europe and resulted in increased literacy and improved scientific communication. (Printing with moveable type of ceramic and wood was invented in China about 1050, by Pi Shêng. Due to the large number of Chinese characters it was not very successful.)

1420–1520

Voyages of exploration, especially by the Portuguese, led to a rapid expansion of geographical and related knowledge.

Medieval optics

1250–1350

A period of heightened interest in 'optics', which at that time included the study of the eye, illusions, perspective, binocular vision, shadows, colours, refraction, the camera obscura, the rainbow, and height of the atmosphere. This interest resulted mainly from Ibn al-Haytham's work of the eleventh century. Some of the leading writers were Dietrich of Freiberg, John of Paris, Kamal al-Din, Roger Bacon, John Pecham, Witelo, and Levi ben Gerson. The latter studied the small variations in the angular diameters of the moon and sun with a camera obscura. Spectacles were mentioned from 1289.

Science in modern times (1500–2000)

The scientific tradition

1600–1700

A new approach to the study of nature (and man) developed in Europe. It was characterized by a mechanistic view of nature and the human body, inspired mainly by the mechanics of Galileo and the philosophy of Descartes. The latter was especially influential in separating answerable from unanswerable questions. This allowed the working of the senses to be investigated with some success, as it led to the realization that perception involves aspects of physics, anatomy, physiology, and psychology.

Francis Bacon's support of empirical science and inductive logic helped to advance the scientific approach. (He did not accept the Copernican theory.)

Scientific institutions began to flourish: the Royal Society of London was founded in 1663, and the Academia del Cimento in Italy some years earlier. The Royal Greenwich Observatory was founded in 1675.

The language of science gradually changed from Latin to various European languages: Newton's *Principia mathematica* was one of the last major works written in Latin.

1700–2000

English, French, German, Italian, and other national languages were used by scientists. This caused communication problems that were almost unheard of in the preceding centuries. During the twentieth century English gradually became the dominant language in science.

The solar system

1543

Copernicus's *On the revolution of the heavenly spheres* contained the hypothesis that the sun, rather than the earth, was the centre of the solar system. Although the idea was not new (see Aristarchus above), Copernicus presented the theory in full mathematical detail. It neatly explained the limited motions of Mercury and Venus and the retrograde motions of the outer planets. The theory was not generally accepted at the time, mainly due to the counterintuitive implication that the earth moves around the sun at great speed. It was not taught to students during the sixteenth century. Copernicus retained the complex system of circular motions and epicycles of the ancients.

1577–1597

A vast series of very accurate planetary observations were made by the Danish astronomer Tycho Brahe. He also inferred from parallax observations that the comet of 1577 was further than the moon (and therefore not an atmospheric phenomenon as Aristotle had thought). Furthermore, the comet's orbit was found to be elongated rather than circular, showing that the planetary spheres have no material existence.

1581–1610

Galileo established the science of mechanics. His work removed the main objection to the Copernican theory, namely that the earth's rapid motion would strip it bare.

1609–1611

Galileo first used a spyglass for astronomical observations. His discovery of the satellites of Jupiter and the phases of Venus contributed to the gradual acceptance of the Copernican system between 1610 and 1640. He also discovered craters on the moon, many stars not visible to the unaided eye, sunspots, and the round appearance of the planets. He explained the dimly visible new moon in terms of earthshine.

1609–1619

Kepler discovered his three laws of planetary motion, confirming the sun as the centre of the solar system. He also did away with celestial spheres and the complex system of circles and epicycles. The importance of his work was not immediately appreciated. The English astronomer Jeremiah Horrocks was the first fully to accept Kepler's laws: in 1640 he applied them to the moon's motion around the earth.

1643

The measurement of atmospheric pressure was initiated by Evangelista Torricelli.

1647

The first lunar atlas was drawn up by Johannes Hevelius, in which the 'seas' and mountain ranges were named.

1651

Giovanni Riccioli named the craters of the moon after famous astronomers, a practice universally adopted since then.

1656–1658

Christiaan Huyghens developed the first accurate pendulum clock, introducing accurate time measurement into astronomy. He also invented a micrometer for the accurate measurement of small angles with the telescope.

1671

The circumference of the earth was measured accurately for the first time, by the French astronomer Jean Picard.

1672

Distances in the solar system were first established with reasonable accuracy by Giovanni Cassini, from his measurements of the parallax of Mars.

1676

The velocity of light was estimated for the first time by Olaus Roemer, from occultations of Jupiter's satellites.

1687

Sir Isaac Newton's *Principia mathematica* revolutionized theoretical astronomy by introducing the theory of universal gravitation and advanced mathematical methods. This work explained Kepler's laws of planetary motion and other phenomena, and predicted many new ones. It was immediately recognized as a momentous achievement.

1781

The planet Uranus, the first new planet to be discovered since antiquity, was identified by William Herschel.

1804

The first scientific observations of the atmosphere at different heights were made from a balloon by Jean Biot and Joseph Gay-Lussac.

1843

The 11-year sunspot cycle was discovered by the German astronomer Heinrich Schwabe.

1846

The planet Neptune was discovered on the basis of perturbations of the orbit of Uranus. The required calculations were made independently by Urbain Leverrier of France and John Adams of England. The discovery provided yet another dramatic demonstration of the power of Newton's work.

1851

The first photographs of the moon were taken by William Bond and Warren de la Rue. A solar eclipse was photographed by the Italian astronomer Pietro Secchi, who also obtained a full set of lunar photographs between 1851 and 1859.

Although the rotation of the earth had been generally accepted since the sixteenth century, the first direct evidence was provided by Jean Foucault's pendulum experiment in 1851.

1861–1868

Spectral analysis was first applied to the sun's light by the Swedish physicist Anders Ångström, and to starlight by Pietro Secchi.

1960–

Space vehicles were utilized to study the solar system, resulting in rapid growth and public awareness of astronomical knowledge.

1969

The first manned moon landing.

Optics and vision

1550–1575

The Italian mathematician Maurolico first explained how the lens of the eye focuses light on the retina, the causes of long- and shortsightedness, and the benefits of concave and convex spectacle lenses. However, his work was published posthumously, after that of Kepler.

1583

The physician Felix Plater recognized the retina as the receptor of visual images.

1604

Johann Kepler asserted, on the basis of his geometric optical analysis of the eye, that the cornea and lens form a real inverted image on the retina.

1621

The Dutchman Snell discovered the sine law of optical refraction. It was first published by Descartes in 1638.

1628

William Harvey described the circulation of the blood. During the next few decades his work helped to end Greek medicine and to found physiology.

1637

René Descartes described accommodation and convergence, and proposed that the latter underlies distance perception.

1668

Mariotte discovered the blind spot.

1709

Bishop George Berkeley distinguished between the angular and perceived size of objects, and explained both size and distance perception in terms of learned cues. These included bodily cues, and visual cues such as aerial perspective. He disowned size–distance invariance.

1729–1760

Pierre Bouguer developed a quantitative optical theory of visibility through the atmosphere. His essay of 1729 contained the main ideas and a more complete theory appeared posthumously in 1760. (The theory is often attributed to Lambert, 1774.)

1789

The astronomer N. Maskelyne calculated the chromatic aberration of the eye and first described night myopia.

1840

Electrophysiology was founded by the German physiologist Emil Du Bois-Reymond, who first detected the small electric currents produced in nerves and muscles.

1850–1870

Substantial developments in sensory physiology and physiological optics were reported by Hermann von Helmholtz.

1860

Psychophysical methods for determining sensory thresholds were developed by Gustav T. Fechner.

1879

Wilhelm Wundt founded the first laboratory devoted entirely to psychological research, including studies of visual perception.

1912

M. Wertheimer founded Gestalt psychology, a school that had a lasting influence on the study of perception.

1917

The term 'size constancy' came into use among Gestalt psychologists.

1950

J. J. Gibson introduced the ecological approach to visual perception, describing perception as entirely stimulus-driven.

1970

The presentation of the visual field on the primary visual cortex, and the specialized visual sensitivities of neurons within it, were described by D. H. Hubel and T. N. Wiesel.

1980–1990

Following studies in visual neuroscience by S. Zeki and others, a broad consensus was reached on three organizing principles of visual analysis: first, analysis is modular, in the sense that different features of a visual image are analysed separately; second, analysis is broadly hierarchical, in the sense that it proceeds in stages; and, third, a considerable degree of parallel processing may occur within stages.

Notes

Chapter 1

1 This term was introduced by Schur (1925) in German and by Holway and Boring (1940a) in English.

2 For example, Angell (1924, 1932).

3 For example, King and Hayes (1966) and Mollon and Ross (1975).

4 Proposed by the Japanese investigator Ryoji Osaka (1962).

5 Claparède (1906a) found the illusion present in children.

6 Cambridge Anthropological Expedition (1901), Vol. 2, pp. 131–2.

7 Some early observations of oval haloes were described by Smith (1738, p. 67) and the subject was later reviewed by Pernter and Exner (1922, Chapter 1).

8 For example, Schlesinger (1913); Minnaert (1954), p. 204; Lynch and Livingston (1995), p. 177.

9 For example, Smith (1738), p. 66.

10 An annotated bibliography of the illusion can be found in Plug (1989a), and a summary of the illusion's early history in Plug and Ross (1989).

11 Euler (1762), Vol. 2, p. 480.

12 Dunn (1762), p. 462.

13 The tablets were studied and translated by R. Campbell Thompson (1900). The quoted passage is from Vol. 2, p. xxxviii.

14 This example was noted by Gregory (1981), p. 102, in his wide-ranging history of explanations in psychology and physics.

15 Ruggles (1999), p. 154.

16 Some authors have suggested that the temple refers to Callanish, and others that it refers to Stonehenge in the south of England. The topic is discussed by Ruggles (1999), pp. 88–9.

17 Diodorus Siculus, *Library of history*, Book II.47.1–5, trans. C. H. Oldfather (1967), Vol. 2, pp. 37–41.

18 The writings of Ibn al-Haytham on optics and vision have been studied by Bauer (1911), Schramm (1959), Lindberg (1967), Omar (1977), Sabra (1989), and others.

19 According to Meyering (1989), the second phase began in the seventeenth century, and the third phase in the nineteenth century; but recent studies of the works of Ptolemy and Ibn al-Haytham show that the second and third phases began much earlier, and overlapped with more primitive views.

20 In Ariotti (1973b).

21 For example, by Cope (1975).

22 The history of this mistake is described by Ross and Ross (1976).

23 Cited by Reimann (1902a), p. 10.

24 Cited by Smith (1905), p. 195.

25 Cited in Reimann (1902a), p. 10.

26 Most authors accept this variability as a fact, though Pozdena (1909, p. 242) thought the enlargement quite steady under widely varying conditions.

27 Euler (1762), Vol. 2, p. 483.

Chapter 2

1 Explanatory Supplement (1961).

2 Weber (1834/1996), p. 125.

3 Kiesow (1925/26), cited by Laming (1986), p. 285.

4 Desaguiliers (1736a), p. 392. Also known as Desaguliers.

5 Explanatory Supplement (1961).

6 Ptolemy, *Almagest*, Book 4, Chapter 6; Dicks (1970), p. 171.

7 Toomer (1974).

8 Ptolemy, *Almagest*, Book 5, Chapters 13–14.

9 Ptolemy, *Almagest*, Book 1, Chapter 3.

10 Cleomedes, *De motu circulari corporum caelestium*, Book 2, Chapter 1. Translated in Ross (2000). An alternative translation appears in Cohen and Drabkin (1948), p. 283.

11 Neugebauer (1975), Vol. 1, p. 103.

12 Lindberg (1970), p. 304.

13 Leonardo da Vinci, *Works*, para. 913.

14 Malebranche (1675), Book 1, Chapter 9.

15 Euler (1762), Vol. 2, p. 481.

16 Quoted in Needham (1959), pp. 225–6.

17 Needham (1959), pp. 226–7.

18 Guthrie (1962), p. 393.

19 Ho Peng Yoke (1966), p. 57.

20 Pannekoek (1961), p. 92.

21 Gubisch (1966).

22 Goldstein (1967), p. 8.

23 Leonardo da Vinci, *Works*, para. 879.

24 Frankel (1978).

25 Haskins, (1960).

26 Wiedemann (1914).

27 Lloyd (1982).

28 Haskins (1960), p. 130.

Chapter 3

1 For example, Hutton (1796, Vol. 2, p. 73) gave the definition that 'apparent magnitude is that which is measured by the optic or visual angle'. Wheatstone (1852) was well aware of the ambiguity of the term, and wrote: 'I do not use the term apparent magnitude, because, according to its ordinary acceptation, it sometimes means what I call retinal, and at other times what I name perceived magnitude.'

2 See Baird (1970) and Poulton (1989) for reviews of the many methods of measurement and the biases associated with them.

3 The effect of instructions was investigated or discussed by many authors, including Gilinsky (1955), Teghtsoonian (1965), Leibowitz and Harvey (1969), Baird (1970), Rock (1977), and Carlson (1977).

4 Smith (1905), p. 195.

5 This was also shown in the study by Teghtsoonian (1965).

6 Failure to recognize each other's definition of perceived size led to a long and acrimonious debate between Malebranche and the physicist Pierre Regis (Malebranche 1675, 1693, 1694; Regis 1694). The nature of the perceived size of the moon was also discussed by Berkeley (1709, Section 74), who took the view that it was the *tangible* rather than the *visual* size that changed – but his definitions do not map clearly onto those of linear and angular size. In more recent times, Witte (1918) showed the important consequences that the choice of definition has on the measurement of the moon illusion, and Kamman (1967) again drew attention to the matter. The theoretical framework for this issue was reviewed by Rock (1975), Hatfield and Epstein (1979), McCready (1986), and Hershenson (1989a, 1989b).

7 Witte (1919b).

8 Aristotle, *De Somniis*, 460b.

9 Fragment 3 in Robinson (1987). The writings of Heraclitus survive only in fragments quoted by later authors. However, the source of this quotation, the philosopher Aetius, lived much later (probably in the first century AD) and his

version of what Heraclitus said may have been influenced by Aristotle's estimate (Robinson 1987, pp. 77–8). The interpretation of this and other classical references to the size of the sun has been studied in detail by Schönbeck (1998).

10 Plug (1989a).

11 Kanda (1933), pp. 293–4.

12 Grant (1974), p. 14.

13 Leonardo da Vinci, *Works*, Vol. 2, p. 120.

14 Talbot (1969), Vol. 3, p. 956.

15 Calculations of perceived distance from linear size estimates formed the basis of von Sterneck's reference level theory of visual space (See Chapters 8 and 9; Sterneck 1907, 1908; Müller 1907, 1918, 1921).

16 For example, Storch (1903) and Witte (1918, 1919c) thought it remarkable that so many authors had studied the horizon enlargement of the moon without first considering why we usually attach a size to it at all.

17 The statements of Epicurus are discussed by Bailey (1926, pp. 61, 125, 287, 391 n15), Siegel (1970), and Rist (1972). Siegel argues that Epicurus and Lucretius held different views, and that Epicurus believed in an incoming effluent stream rather than particles. Epicurus wrote in a letter to Herodotus: 'we see and recognize by the entrance into the eye of something from the outside … of similar color and shape and also of a size rendered suitable for entry into the visual organ and the mind' (Siegel 1970, p. 20). This statement suggests that there was a perspective diminution of the image as it reached the eye, the image being scaled in size to fit into the pupil. The question of how it could be rescaled to give true size was not discussed.

18 Translated by C. Bailey (1947), pp. 188–9.

19 Cleomedes' dates are controversial, but Bowen and Goldstein (1996, 171 n27) argue that he lived after Ptolemy because he refers to the Equation of Time, a concept little known before Ptolemy. See also Bowen and Todd (in press) for a translation and commentary on Cleomedes' book on 'The Heavens'.

20 Blumenthal (1971).

21 For historical reviews see Boring (1942) and Ross and Plug (1998).

22 Translated by MacKenna and Page (1952), p. 65.

23 See review by Epstein, Park, and Casey (1961).

24 Translated by Burton (1945), p. 358.

25 See Smith (1982). In yet another passage (Theorem 57) there is an added note that the perceived sizes of equally large objects at unequal distances are not inversely proportional to their distances: the author of this note was therefore aware of the distinction between visual angle and perceived size (Boring 1942); but since the authenticity of this passage is doubtful we cannot be sure that the distinction goes back that far (Lindberg 1976).

26 Ptolemy's *Optics* exists only as a twelfth-century translation by Eugene of Sicily into medieval Latin from an Arabic translation of the Greek original: however, most of it is thought to be Ptolemy's own work rather than a later Arabic addition (Sabra 1987; Smith 1996).

27 *Optics*, II, 53–63, and several other sections.

28 Our translation.

29 Ross (2000).

30 Our translation from the Greek edition of Todd (1990), p. 46.

31 Balfour was a Roman Catholic Scot from St Andrews, who later held chairs in philosophy and mathematics at the college of Guyenne in Bordeaux, where he produced a Greek and Latin edition of Cleomedes.

32 Ibn al-Haytham credited the passage to Ptolemy's *Optics*, thus showing that this material was not a twelfth-century addition (Sabra 1987).

33 Translated from the Latin in Reimann (1902a), p. 2.

34 See review of the early constancy literature by Epstein (1977).

35 For example, Joynson (1949), Ittelson (1951), Rock and McDermott (1964).

36 The sketches of Cornish (1937) are an example of this method (Chapter 4).

37 There seem to be no early examples of these techniques, but Higashiyama (1992) used both methods and found compatible results with each.

38 For reviews of the size constancy literature see Epstein, Park, and Casey (1961), Gogel (1977), Carlson (1977), McKee and Smallman (1998). Detailed discussions of the nature of size constancy can be found in the textbooks of Kaufman (1974) and Rock (1975). Baird (1970) gives a psychophysical analysis of many experiments.

39 For example, Witte (1918) and Epstein (1977).

40. This difficulty was pointed out by Joynson (1949) and later authors.

41 For example, Rock and McDermott (1964) argue that in reduced cue conditions visual angle or retinal size directly evokes perceived size, without the observer making any assumptions about equidistance.

42 For example, Lichten and Lurie (1950), Rock and McDermott (1964). Baird (1970) reviews these and other experiments in which angular size is correctly matched for targets presented in reduced cue conditions. Wallach and McKenna (1960) argued that angular size could not be correctly perceived, on the grounds that the majority of their observers gave different matches on ascending and descending trials: however, they did not publish group means and standard deviations, and seem only to have shown that angular matches can be imprecise under some test procedures.

43 Holway and Boring (1941) displayed a standard target of variable linear size but constant angular size (1 degree) at distances of up to 100 ft (30 m), and asked observers to match it in 'perceived size' by a variable target at 10 ft viewed at right angles to the standard. The authors presented their results graphically, showing the

linear size matched at different distances, so that the function for a visual angle match had a slope of zero and that for a linear match had a slope of 1.0. The lowest slope (0.22) was found in the most 'reduced cue' condition, and the slope increased with the availability of more cues. This manner of presenting the results makes it difficult to appreciate the extent of the angular enlargement with distance. The viewing distances were 10, 50, and 100 ft, and the mean matched sizes were 2.2, 3.9, and 6.4 inches respectively, equivalent to angular matches of 1.05, 1.86, and 3.05 degrees. The linear size match required for an angular match of 1 degree was always 2.09 inches. For these small angles the linear size ratios are equivalent to the angular size ratios.

44 The two observers were A. H. Holway and A. C. S. Holway. According to Gilinsky (1955, p. 179), Boring confirmed in a personal communication that he intended observers to make a retinal match.

45 Gilinsky (1955) presented four standard targets, triangles of heights from 42 to 78 inches (1.07–1.98 m), at distances of 100–4000 ft (30–1220 m) in an open field. Thirty-two to 36 observers matched them in perceived 'retinal' size by an adjustable target at 100 ft, located at 36 degrees to the right of the standard targets. (More recently Gilinsky (1989) stated that her 1955 data represented 'picture image' matches rather than visual angle matches, and that the latter would be smaller.) Gilinsky presented the mean results graphically in two ways: as a percentage of the matched linear size to the true linear size as a function of distance; and transformed to a constant visual angle format for comparison with the graphs of Holway and Boring (1941). Neither of these formats reveal the true extent of the overestimation of angular size. We calculated the mean linear size matched for a mean standard of 60 inches at each distance from the data in Gilinsky's Table 1, and found the corresponding angular ratios. This gives the following table:

Distance of standard (ft)	100	200	400	800	1600	4000
Matched size (in)	58.5	45.8	34.1	20.5	12.5	6.1
Angle of match (degrees)	2.79	2.19	1.63	0.98	0.60	0.29
Angle of standard (degrees)	2.86	1.43	0.72	0.36	0.18	0.07
Ratio of angles	0.98	1.53	2.26	2.72	3.33	4.14

46 The results shown here are the mean data from experiment 1 of Leibowitz and Harvey (1969) with retinal match instructions, and the similar experiment from Leibowitz and Harvey (1967). (These results are shown together in the 1969 paper, Fig. 1). Twenty-three observers contributed to the 1967 experiment and 20 to the 1969 experiment. The observers matched the perceived length of an adjustable rod at 51 ft to that of a 6 ft standard target (a board alone, or a board with a person beside it, or a person alone) viewed in a university mall at distances of 340 to 1680 ft. The rod was located at 90 degrees to the left of the target. The authors presented their results in the form of Brunswik constancy ratios. Brunswik ratios are a particularly misleading way of presenting results from size matches, because the ratios change dramatically if the values for the standard and adjustable targets

are interchanged (Sedgwick 1986). The ratio is $(S'-s)/(S-s)$, where all values are linear: S is the standard target, S' is the matched adjustable target, and s is the linear size of a true angular match to the standard target. To calculate the angular size ratios, we had to estimate the value of the Brunswik ratios from the published graphs, insert the values for the known variables, and calculate the missing values. The calculated values, to which we have added the notional value of 1.0 at 51 ft, are given in the following table:

Distance (ft)	51	340	680	1020	1360	1680
Brunswik ratio		0.33	0.30	0.25	0.24	0.20
Angle of standard (degrees)		1.01	0.51	0.34	0.25	0.20
s (ft)		0.90	0.45	0.30	0.22	0.18
S' (ft)		2.58	2.12	1.77	1.63	1.36
Ratio of angles	(1.0)	2.87	4.71	5.90	7.41	7.56

47 One of the causes in some experiments may be the 'error of the standard' – the tendency to judge the standard as too large, or (equivalently) to set the matched variable too large, even when both targets are at the same distance. The further target was always the standard in these experiments. However, if the size of the error is constant, it cannot explain the growth with distance. Holway and Boring (1941) describe it as a 'place error', but found an effect of only about 5 per cent in the data we have shown here. Discussions of the error can also be found in Piaget and Lambercier (1943) and Lambercier (1946). Baird (1970, p. 82) disputes whether this error is commonly found in angular matching experiments, and indeed it is not present in the data of Gilinsky (1955). Leibowitz and Harvey (1967, 1969) did not match the variable and standard at the same viewing distance.

48 Higashiyama (1992) did not report his results in this form. Instead he noted that there was a hyperbolic relation between judged and true angular size, with small angles being overestimated more than large angles. He gave this latter finding as an explanation of size constancy (Chapter 10). We calculated that his data for a 10 degree target, combining vertical and horizontal target orientations, showed overestimation by a factor of about 1.2 at 3 m, 1.7 at 10 m, and 1.9 at 30 m.

49 For several years one of the authors (HER) asked a class of psychology students to estimate the extent to which distant hills appear larger in real life than they do in photographs, and obtained median values of 3 or 4, with a range from 2 to 8. The interpretation of size in photographs is discussed in Chapter 4.

50 Leibowitz and Harvey believed that they were measuring perceived retinal size, and speculated that the overestimation of retinal size might contribute to the overestimation of linear size for unfamiliar objects. They thus subscribed to some form of perceptual size–distance invariance, in which angular size is misperceived.

51 A fundamental difficulty with both classical and perceptual size–distance invariance is that they are based on a geometrical model in which 'linear size' refers to the height of an object at right angles to the ground, and 'distance' refers to length or

size along the ground. In practice, the observer's eye is above the ground, and both ground and object may slope. Both the ground size and the object size subtend angles at the observer's eye, and all the linear and angular measures are linked: there is no reason to suppose that any two measures are primary and determine the third, and it is unclear what is to be taken as the distance to a sloping object (Schwartz 1994). We describe some more complex geometrical models involving slope in Chapter 9.

52 Other twentieth-century approaches to size and distance perception include Gestalt theory, Brunswik's 'probabilistic functionalism', and J. J. Gibson's 'direct perception'. A readable account of these, and of some neurophysiological approaches, can be found in Gordon (1997). Lombardo (1987) and Schwartz (1994) provide more detailed evaluations of Berkeley and Gibson and their background.

53 Mayr (1904) and Claparède (1906a) discussed about a dozen, and the creation of new ones (or re-creation of old ones in slightly modified form) has not yet abated. Osaka (1962) distinguished 18 theories, indicating the main proponents and critics of each.

54 Comprehensive reviews of the geometrical illusions can be found in Robinson (1972) and Coren and Girgus (1978). Gillam (1998) reviews more recent work on some geometrical illusions. Gregory (1963) takes the view that many size illusions are the result of inappropriate size constancy. He also (1998, pp. 246–52) gives a broader classification of different types of illusions, as 'ambiguities, distortions, paradoxes and fictions', each of which may have several causes.

55 For example, Descartes argued that the application of *natural geometry* in spatial perception was innate, at least for the perception of distance from the convergence angle of the two eyes. His description of size–distance invariance, however, seems to rely more on cognitive processes. The history of the idea of natural geometry is described by Pastore (1971) and Cutting (1986).

Chapter 4

1 Witte (1919a) commented on this methodological problem.

2 One of the Canary Islands, at a latitude of about 28 degrees north.

3 Kaufman and Rock (1962a, 1962b); Rock and Kaufman (1962).

4 Poulton (1989).

5 A very readable discussion of the meaning of size in art can be found in Gregory (1998).

6 Craddick (1963).

7 For example, Edgerton (1975), Lindberg (1976), Kemp (1990), and Wade (1998).

8 Euclid never specified that parallel lines actually converge to a vanishing point in the distance, perhaps because this seemed to conflict with the concept of the 'visual cone' with outgoing rays diverging from the eye. The perspective cone and the visual

cone appear to be the reverse of each other, a point that also troubled Leonardo da Vinci in the fifteenth century:

> Perspective employs two opposite pyramids [i.e. cones]: one of which has its apex in the eye and its base far away at the horizon; the other has its base at the eye and its apex on the horizon. The first is embracing all the objects that pass in front of the eye; the second has to do with the peculiarity of the landscape, in showing itself so much smaller in proportion as it recedes farther from the eye (abbreviated by Siegel 1970, p. 102).

This point has been known to trouble today's students. However, it can be resolved by remembering that Euclid was describing space in terms of the visual angle subtended at the eye, while the perspectivists were showing how the visual angles translated into lengths on the picture plane.

9 Translated by Burton (1945).

10 Willats (1997), pp. 59–61.

11 Claparède (1906a), Kaufman and Rock (1962b), Ross, Jenkins, and Johnstone (1980).

12 Willats (1997), p. 168.

Chapter 5

1 For example, Haenel (1909) and Lohmann (1920).

2 Aristotle, *Meteorologica*, Book 3, Chapter 4, 373b,12–13, trans. Lee (1962).

3 Aristotle, *De sensu*, Chapter 2.

4 Edelstein and Kidd (1972).

5 Strabo, *The geography*, Book 3, Chapter 1. Our translation.

6 Seneca, *Naturales quaestiones* (I, 6), trans. Corcoran (1971).

7 Ptolemy, *The Almagest*, Book 1, Chapter 3, trans. Ross and Ross (1976), p. 378.

8 Cleomedes, *Meteora*, Book 2, Chapter 1, trans. Ross (2000).

9 Ptolemy, *Optics*, V, 77, trans. Smith (1996), p. 257.

10 Sabra (1987, 1996).

11 Ross and Ross (1976); Sabra (1987, 1996).

12 Sabra (1989), Vol. II, p. xxxv.

13 Sabra (1996).

14 Sabra (1996), p. 38.

15 Trans. Sabra (1989), Vol. 1, pp. 340–1.

16 Macrobius, *The Saturnalia*, Book 7, Chapter 14.2, trans. Davies (1969).

17 Robins (1761), Vol. 2, p. 235.

18 Grant (1974), pp. 380–3.

19 Grant (1974), p. 444.

20 For example, Roger Bacon (1263), John Pecham (1274), and Mario Bettini (1642).

21 Birch (1757), Vol. 4, p. 240.

22 Ariotti (1973b), p. 15.

23 Birch (1756), Vol. 1, p. 491.

24 Birch (1757), Vol. 4, p. 530.

25 Birch (1757), Vol. 3, p. 503.

26 Hobbes (1658), Vol. 1, p. 462.

27 Kinsler (1945).

28 Sears (1958); Ross (1970).

29 Ross (1967) found that linear size in water was overestimated by an average factor of 1.18 compared to estimates in air, and the optical distance by 1.25. Other experiments are summarized by Ross and Plug (1998). Optical magnification in water does not seem to offer a straightforward explanation for the perceived values of either angular or linear size.

30 Ross (1967).

31 Thouless (1968) concluded that size–distance invariance did not hold, which is strange since he seems to have measured perceived angular size rather than perceived linear size.

32 Explanatory Supplement (1961).

33 Pannekoek (1961).

34 Thorndike (1923–58), Vol. 5.

35 Hirschberg (1898).

36 Photographs of the flattened sun have been published by Mollon and Ross (1975) and Walker (1978a), in connection with the celestial illusion. A photograph of the flattened moon can be found in Lynch and Livingston (1995), p. 46.

37 Ferguson (1999).

38 Birch (1757), Vol. 3, p. 183.

39 Mollon and Ross (1975).

40 Birch (1757), Vol. 3, pp. 98, 306–8.

41 Birch (1757), Vol. 3, p. 183.

42 Turning one's head through 90 degrees onto the left or right shoulder makes it much more noticeable. This is presumably because of the horizontal–vertical illusion – the perceived elongation of vertical lines in the retinal image.

43 Robins (1761), p. 242.

44 Photographs of mirages can be found in Lynch and Livingston (1995), pp. 52–8.

45 This curvature illusion was described by Titchener (1901), Luckiesh (1922/65, p. 3), Tolansky (1964, pp. 1–54), and several later authors.

Chapter 6

1 Middleton (1960).

2 Middleton (1958).

3 Fry, Bridgman, and Ellerbrock (1949) showed that low luminance contrast increased perceived distance in a stereoscopic range-finding experiment, and argued that contrast had a specific effect on the binocular stereoscopic mechanism. However, Ross (1971, 1993) conducted a monocular experiment and found a regular relation between the numerically estimated distance of objects viewed through murky water and their luminance contrast. O'Shea, Blackburn, and Ono (1994) also found that lower-contrast targets appeared further away in pictorial displays, the effect being greater with monocular than binocular vision. The effect on perceived distance does not, therefore, depend on binocular vision.

4 Ptolemy, *Optics*, II, 124, trans. Smith (1996), p. 120.

5 Boring (1942).

6 Siegel (1970), p. 86.

7 Siegel (1970), p. 86.

8 Plotinus, *Second Ennead*, VIII, trans. S. MacKenna and B. S. Page, pp. 64–5.

9 Cleomedes, *Meteora*, Book 2, Chapter 1, trans. Ross (2000).

10 Ptolemy, *Optics*, II, 126, trans. Smith (1996), p. 121.

11 Leonardo da Vinci, *Notebooks*, Vol. 1, pp. 262–3.

12 Our translation.

13 Hamilton (1849), p. 191.

14 Tomlinson (1862), pp. 309-10.

15 Helmholtz (1856–66/1962), Vol. 3, p. 283.

16 Ibn al-Haytham, *Optics*, III.7, para. 194, trans. Sabra (1989), Vol. 1, pp. 340–1.

17 The impoverished visual scene in a fog may also prevent walkers from compensating adequately for their own motion, with the result that they perceive nearby rocks as moving towards or past themselves. This may give rise to the illusion that the rocks are people walking around (Ross 1994b).

18 Kaufman (1979, pp. 220–6) gives a readable analysis of the change of image size for close objects during locomotion, and the use of this cue to judge the true distance of the object.

19 An uninformative experiment was conducted by Pickford (1943), who examined the effects of 'veiling glare' on the perceived size and distance of two white discs located inside a box and viewed binocularly through two peepholes. The two discs differed in the amount of glare they received from lamps, and their distance was altered so that they appeared equal either in distance or in perceived (angular) size. The results for the 9–11 observers were erratic. Pickford concluded that 'apparent size and distance become separate factors in fog vision', although he thought that

perfect (linear) size constancy held for normal viewing conditions. It is unclear how he reached these conclusions, because veiling glare does not imitate a true fog, and the method of matching angular size does not measure perceived linear size.

20 Aerial perspective was mentioned as a contributory cause of the illusion by Porterfield (1759), Weber (1846), Helmholtz (1856–66), Wundt (1873), James (1890), and more recently by Furlong (1972).

21 Euler (1762), Vol. 2, letters 113 and 114.

22 For example, Eginitis (1898) and Claparède (1906a). Brewster (1849/1983) also claimed that circles viewed in depth through a stereoscope did not change in perceived size when their brightness was changed, and that this counted against Berkeley's explanation of the moon illusion.

23 Berkeley (1709), Section 72.

24 This observation was supported by Lewis (1898) and Angell (1924), who used it to argue that aerial perspective could not explain the sun illusion, and could explain the moon illusion only when it occurred during daylight or twilight.

25 Plato wrote in the *Timaeus* (67d–e): 'The particles that come from other bodies and enter the visual ray when they encounter it, are sometimes smaller, sometimes larger than those of the visual ray itself – The names should be assigned accordingly: 'white' to what dilates the visual ray, 'black' to what contracts it' (trans. Cornford 1959, p. 80).

26 Ptolemy, *Optics*, II, 24, trans. Smith (1966), p. 81.

27 Ibn al-Haytham, *Optics*, III.7, paras. 250–3, trans. Sabra (1989), Vol. I, p. 357.

28 Leonardo da Vinci, *Works*, Vol. 1, p. 214.

29 Fletcher (1963).

30 Descartes, *Dioptrics*, Discourse VI (1637a, p. 255).

31 Descartes (1637b), trans. Olscamp (1965), pp. 111–12.

32 Descartes (1637a), p. 255.

33 Helmholtz (1856–66), Vol. 2, pp. 186–93, reviewed irradiation, but it had already been studied extensively by Plateau (1839a, 1839b).

34 See Farnè (1977) and Weale (1975).

35 Over (1962b).

36 For example, Ashley (1898), Egusa (1983), Farnè (1977).

37 Astronomers measure the brightness of celestial bodies on a stellar magnitude scale, whose origins date back to Hipparchus in the second century BC. This scale is logarithmic, and gives a value of 6 to the faintest stars visible to the naked eye, and low or negative values to brighter objects. The full moon is the brightest celestial object visible from earth in the night sky. On the stellar magnitude scale it has a value of about −12.7, while the next brightest object is the planet Venus which reaches a maximum of −4.4. The unobscured sun has a value of −26.8, and is too bright to

look at without damaging the eye (Ridpath 1998, pp. 132–3; Zeilik and Gregory 1998, pp. 226–7).

38 It was usually the position of the horns of the crescent moon that were thought to predict the weather. In many cultures bad weather was predicted when the horns point upwards, and fair weather when they point horizontally (Wallis 1999, p. 301). Note also the various events predicted by the moon's appearance, quoted in Chapter 1 from an early Assyrian tablet.

39 Quiller-Couch (1955), p. 329.

40 Strous (1999) and Conrad (2000).

41 In terms of stellar magnitude, the moon reduces from about −12.7 when full to about −5.2 when almost new (Allen 1973, p. 143).

42 Conrad (2000).

43 Lynch and Livingston (1995), p. 202.

44 A minor variable affecting the moon's brightness is its distance from the sun and from the earth. The light falling on the moon varies very little with its distance from the sun, but the light reflected back to us is about 30 per cent brighter for the closest than for the furthest distance to the earth (Walker 1997). This can produce an increase of about 25 per cent in total brightness for the close moon compared to a moon of average distance, when brightness is summed over the whole disc of the moon (which is angularly enlarged by about 5.5 per cent, as described in Chapter 2). Brightness also varies with the seasons, the average winter solstice midnight full moon being about 10 per cent brighter than the summer equivalent (Strous 1999).

45 Goethe (1810), pp. 29–30.

46 Minnaert (1954), pp. 104–5.

47 Humphreys (1964), pp. 557–62.

48 For example, Müller (1906), Haenel (1909), Leiri (1931), Furlong (1972).

49 Henning (1919), p. 277, our translation.

50 Taylor and Sumner (1945), Johns and Sumner (1948), Payne (1964).

51 Luckiesh (1918), Pillsbury and Schaefer (1937).

52 Ross (1975).

53 Simonet and Campbell (1990). The explanations have been reviewed by Sundet (1972, 1978) and Ye, Bradley, Thibos, and Zhang (1991).

54 Oyama and Yamamura (1960), Kishto (1965), Sundet (1972).

55 Kaufman and Rock (1989), pp. 212–13.

56 Kaufman (1960), pp. 152–3.

Chapter 7

1 Gassendi (1636–42), pp. 421–2.

2 Lindberg (1976), p. 42.

3 Porterfield (1759), Vol. 2, pp. 92–4.

4 Leonardo da Vinci (1452–1519), *Works*, para. 31; see also paras 32 and 38; *Notebooks*, pp. 211, 238; *Commentary*, Vol. 1, p. 117.

5 See Lindberg (1976) and Wade (1998a).

6 Gassendi (1636–42), pp. 422–48.

7 Birch (1756–7), Vol. 3, pp. 502–3.

8 Enright (1975).

9 Porterfield (1759), Vol. 2, p. 381.

10 Walker (1804, 1805, 1806, 1807).

11 Mayr (1904).

12 Leonardo da Vinci (1452–19), *Notebooks*, pp. 238–9.

13 For example, Young (1807), Reimann (1902a), Mayr (1904).

14 Porterfield (1759), Vol. 2, p. 181.

15 Barlow and Mollon (1982), p. 56.

16 Gubisch (1966).

17 Barlow and Mollon (1982), p. 58.

18 The data on chromatic aberration was quoted by Leiri (1931) from Helmholtz's authoritative review in 1867.

19 For example, Knoll (1952), Coren and Girgus (1978).

20 Barlow and Mollon (1982), p. 59.

21 Knoll (1952), Levene (1965).

22 Otero (1951), Campbell (1954), Owens (1979), Leibowitz and Owens (1975).

23 Mellerio (1966).

24 Mellerio (1966), Leibowitz, Hennessy, and Owens (1975).

25 Boff and Lincoln (1988), Vol. 1, p. 114.

26 Meehan and Day (1995).

27 Heinemann, Tulving, and Nachmias (1959) presented a standard circle of light subtending 1 degree at the nodal point of the eye at distances between 25 and 300 cm. Observers decided whether a variable target at 400 cm was larger or smaller, or further or nearer, than the standard. The point of subjective equality (PSE) for the size judgements was calculated. The mean PSE reduced from about 70 mm at 300 cm to about 57 mm at 25 cm, showing a change of about 20 per cent. The results were the same for 12 observers who had normal accommodation, and for four observers who had their accommodation paralysed by homatropine and their pupil size controlled by a 0.05 mm artificial pupil. Convergence therefore seemed to be the main cause of the changes in perceived size. However, the reported size of the effect may be too large, because of a possible 'error of the standard' – the distance of the standard and variable was not counterbalanced.

28 Recent discussions of oculomotor micropsia are given by Ono, Muter, and Mitson (1974), Alexander (1975), McCready (1985, 1999), and Meehan and Day (1995).

29 Fry (1983).

30 Similar results were found when the experiment was repeated by Simonelli and Roscoe (1979), both during the day and at night. Earlier studies at the Ames research centre of the National Aeronautics and Space Administration had also yielded a strong relationship between judgements of perceived size and focusing distance (Hull, Gill, and Roscoe 1982).

31 Wolbarsht and Lockhead (1985).

32 Kaufman and Rock (1989), Lockhead and Wolbarsht (1989).

33 Benel (1979), Roscoe (1982, 1985).

Chapter 8

1 For example, Tscherning (1904), p. 315.

2 Reimann (1902a), pp. 2–3, quotes parts of the Latin translation given in Risner's 1572 edition, *Opticae thesaurus Alhazeni*, Book 7, Section 55.

3 Gray (1964), Vol. 8, p. 308.

4 Kolb (1999), p. 274.

5 Gray (1964), Vol. 2, p. 335.

6 Old Testament, Isaiah 40.

7 Gray (1964), Vol. 12, pp. 34–5.

8 Gray (1964), Vol. 8, pp. 333–7.

9 Gray (1964), Vol. 8, p. 221.

10 Jackson (1971), p. 80.

11 Warner (1996).

12 Trans. Schönbeck (1998), p. 34.

13 Grant (1974), p. 444.

14 Treiber's Latin treatise on the form and colour of the sky was reviewed by Zehender (1900).

15 Robins (1761), Vol. 2, pp. 240–1.

16 Filehne (1894).

17 For example, Zoth (1902), Mayr (1904), Reimann (1904), Haenel (1909), Müller (1921), Pernter and Exner (1922), Filehne (1923).

18 Angell (1932), p. 135.

19 Zehender (1900).

20 Zehender (1900).

21 For example, Robins (1761), Dember and Uibe (1918, 1920), Pernter and Exner (1922).

22 Martin Folkes (1690–1754) was an English antiquarian and Fellow of the Royal Society of London.

23 Reimann (1902a).

24 Zehender (1900).

25 Minnaert (1954), p. 164.

26 Ariotti (1973a, 1973c). Castelli himself did not give a perceived distance explanation of the perceived size of afterimages. Instead he gave a relative size explanation, as he did for size constancy and the celestial illusion (Chapter 10). On this account the more distant afterimage appears larger because its angular size has increased relative to that of adjacent objects.

27 Boring (1942), when outlining the history of Emmert's law, used several terms interchangeably; but it is clear that he intended the law to refer to perceived rather than true size and distance, because he believed the law to be a special case of size constancy or size–distance invariance (Boring 1940). Emmert himself seems to have been unaware of a possible distinction between perceived and physical size (Epstein, Park, and Casey 1961).

28 Experimental results depend very much on the method of measurement of the size of the afterimage. Indirect comparisons with other objects tend to produce results dependent on perceived distance, whereas direct measurement by markers more nearly follows physical distance (Epstein *et al.* 1961).

29 Filehne (1894) confirmed that an afterimage of the sun appears smaller when raising one's eyes to the zenith, and attributed its discovery to Ewald Hering. Zoth (1899) used both the sun and an electric arc viewed through various coloured filters to create afterimages of different angular sizes and colours; all of these appeared larger against the horizon sky than overhead, but Zoth could not say whether they appeared closer or further. Reimann (1902b) and Mayr (1904) reported similar results, Mayr adding that occasionally an afterimage of the sun would appear to be on the horizon, rather than floating in front of it, and would then seem very large. Pernter, with a slightly different procedure, showed a similar effect: he formed an afterimage of the sun when high in the sky and brought it down against the horizon sky, where it appeared much enlarged (Pernter and Exner 1922, p. 46).

30 For example, Hobbes (1658), Zeno (1862), the astronomers Biot and Bohnenberger, Reimann (1902b, 1904), Dember and Uibe (1918, 1920).

31 Pernter and Exner (1922), p. 6

32 Smith (1738), Section 162, p. 63.

33 J. C. E. Schmidt (1834), *Lehrbuch der analytischen Optik*, cited by Reimann (1902a).

34 Helmholtz (1856–66), pp. 290–2.

35 Euler (1762), Zeno (1862), Reimann (1902b), and Dember and Uibe (1920).

36 Plug (1989a).

37 Minnaert (1954), pp. 165–6; Schönbeck (1998).

22 Higashiyama's definition bears some similarities to McCready's definition of angular size as the difference in angular direction between two end-points.

Chapter 11

1 Trans. Granger (1970), Vol. 1, p. 191.

2 The Latin text and a French translation can be found in Lejeune (1989), pp. 115–16. The translation given here is by Ross and Ross (1976). An alternative English translation, and a further discussion of the text, is available in Smith (1996), p. 151.

3 Zoth (1902) pointed out the difficulties in interpreting this passage from Ptolemy. He suggested that it was not unambiguously an angle of regard theory, as Reimann (1902a) had claimed. It might refer to size–distance invariance, or perhaps to the use of peripheral vision. Zoth suggested that by 'viewing under normal conditions' Ptolemy may have meant direct or central vision, and by 'unusual conditions and the difficulty of the action' indirect or peripheral vision. We discuss the peripheral vision explanation later in the chapter.

4 For example, Lejeune (1989), p. 116.

5 *Planetary hypotheses*, Book I, Part 2, Section 7. This passage is translated by Goldstein (1967) and by Sabra (1987), who discuss its ambiguities.

6 The idea that Ptolemy ascribed the moon illusion to intervening objects probably stems from Della Porta in his *De refractione* (1593). The history of these later ascriptions is described by Ross and Ross (1976).

7 O'Neil (1957, 1969) argued that Ptolemy's main concern was to explain the motions of the planets (which included the sun and the moon). Ptolemy followed Hipparchus in believing that the planets followed circular paths with uniform velocity, but with centres offset from the position of the earth. The main circular path was known as the deferent. In addition, the planets followed smaller circles, or epicycles, whose centre moved around the circumference of the deferent. The right combinations of offsets, deferents, and epicycles could be used to account in geometrical terms for the irregular paths and velocities of the planets as observed from the earth. Ptolemy's theory of the moon's orbit implies large variations in the angular size of the moon during each month, but of a different type from the horizon–zenith illusion. Ptolemy may have been anxious to dispose of the latter illusion because it confused the issue for his planetary theory, and he may have produced more than one explanation to make sure of dismissing the effect. O'Neil concluded that Ptolemy held some psychological explanation of the illusion: he followed Boring (1943) in the incorrect assumption that Ptolemy used size–distance invariance combined with intervening objects.

8 Gauss (1880), pp. 498–9. Our translation.

9 Negative results were found by Zoth (1899, 1902), Mayr (1904), and Dember and Uibe (1920).

10 The third observer was Leo M. Hurvich, famous for his work on brightness and colour vision.

11 In addition to the experiments described in the text, positive results were found by Zoth (1899) and Leibowitz and Hartman (1959a).

12 Bilderback, Taylor, and Thor (1964) produced the artificial moons by illuminated apertures 2.8 cm wide, which were displayed at a distance of 2.7 m from the observer's eyes. The mean matched size was 23.2 cm for the horizon moon and 21.9 cm for the zenith moon. The authors did not state the distance of the matched circles, but these large sizes suggest that the distance was much greater than that of the moons. The mean estimated distances were 8.21 and 10.33 ft respectively (2.5 and 3.2 m).

13 Thor, Winters, and Hoats (1970) had the children compare standard and variable illuminated circles which they viewed through collimating lenses in a dark room. The lens-to-eye distance was 1.52 m. The targets were variable in size, with diameters of 0.99 to 3.35 mm. The children were required to make size adjustments so that the two targets appeared equal in either size or distance. With the lower target as the standard, the elevated target appeared smaller and further, but when the higher target was the standard there was no illusion. The ratio of upper to lower size settings (for elevations of 90 and 10 degrees) was 0.95 when the standard was high and 1.19 when the standard was low. The authors concluded that the 'error of the standard' might account for about half of the effect obtained in previous experiments on eye elevation. The combined results for both positions of the standard gives a size illusion of about 7 per cent.

14 The 'error of the standard' normally takes the form of adjusting the variable to be larger than the standard. Thus when the standard is in the zenith the measured illusion should be less than when the standard is on the horizon. Without a counterbalanced design, we cannot be sure that this was the direction of the error in Suzuki's experiment. A good discussion of the error of the standard in eye elevation experiments can be found in Kaufman and Rock (1989).

15 Barlow and Mollon (1982), p. 63.

16 See review by Ross (1997).

17 It should be noted that binocular disparity usually gives a positive correlation between changes in perceived size and distance. However, oculomotor micropsia produces a decrease in perceived size without a corresponding decrease in perceived distance: it can therefore be regarded primarily as a decrease in perceived angular size.

18 McCready (1986) suggested that the purpose of oculomotor micropsia is to compensate for the fact that the centre of rotation of the head lies about 10–15 cm behind that of the eyes: if uncorrected, this should cause a discrepancy between head and eye alignment for close objects. Minification of the perceived visual angle of close objects should thus help in producing an appropriate linkage between head and eye movements.

Chapter 12

1 See reviews by Scheerer (1984, 1987).

2 The sensory aspect of sensorimotor systems is sometimes referred to as *kinaesthesis* (particularly when the sense of movement is stressed), and sometimes as *proprioception* (when static bodily information is included).

3 Wade (1998b).

4 Boring (1942).

5 Wood, Zinkus, and Mountjoy (1968) mentioned recent discoveries of neural coordination between sensory fibres of the vestibular system and the visual cortex, as described by Jung (1961). It is curious that the existence of such neural connections should be considered as an explanation of a vestibular effect on size perception. The connections that Jung described were concerned with the effect of angular acceleration on reflex eye movements. What needs to be shown is that there is a neural connection between otolithic input and some part of the visual system that analyses size.

6 In the first experiment of Van de Geer and Zwaan (1964), 26 observers adopted an upright or a supine posture, and viewed luminous circles either with horizontal gaze or with vertical gaze (head tilted back and eyes raised maximally). The viewing distance of the targets varied between 3 and 45 m, and the linear size of the vertical target varied with the distance so that it always subtended half a degree. The observer was required to vary the size of the horizontal target to match that of the vertical target. (The authors report these as linear size matches, but presumably they were angular matches since size and distance covaried, and the experiment was conducted in the dark.) In the supine posture, the observers made accurate matches for all sizes and viewing distances. In the upright posture, the horizontal targets were adjusted to be smaller than the vertical targets, showing that their perceived size was relatively enlarged.

7 The observers adjusted the size of a variable circle at distances of 5–40 m so that it appeared equal to a 10 cm diameter circle at a distance of 3 m. The matched size of the further target was always smaller than that required for an angular match, showing perceived enlargement with viewing distance.

8 Four observers adjusted a circle viewed with eyes raised 60 degrees to appear equal in size to a 45 cm circle viewed horizontally. The viewing distance was 40 m. With binocular vision and upright posture the raised circle was adjusted to 56 cm, giving a diminution effect of 24 per cent; while with monocular vision it was adjusted to 53.25 cm, giving a diminution effect of 18 per cent. With supine posture and raised eyes there was no diminution for either binocular or monocular vision. Interestingly, after six hours of monocular vision, the vertical diminution had increased to 26 per cent for upright posture and 18 per cent for supine posture. This may suggest that some perceptual learning occurs, but it contradicts the finding of Taylor and Boring (1942) that prolonged monocular viewing reduces the illusion.

9 Ross (1965) used a 'trombone' apparatus, which contained a white disc of 12.7 cm diameter fixed at a distance 33 cm from the viewing point, and a movable disc of 14 cm diameter. The observer adjusted the distance of the larger disc so that the two discs appeared equal in angular size. The largest distance settings (or greatest angular size enlargement) occurred for the following order of combinations: apparatus horizontal, body upright; apparatus horizontal, body supine; apparatus vertical, body upright; and lastly apparatus vertical, body supine. However, the method of measurement confounds perceived angular size and perceived distance, and it is impossible to compare perceived size across orientations from measurements taken within orientations.

10 Ross (1965) also tested five divers under water with the same experimental design. Unlike the equivalent land experiment, there were no differences in the settings for any combinations of target and diver orientation. The most likely cause was the low visibility (about 8 m), which removed any differences in the visual background for different target orientations: a reduction in perceived size on looking up to the surface does occur in clear water (Ross, King, and Snowden 1970). The reduction in tactile–kinaesthetic information in water may also have reduced any effect of diver posture.

11 Zinkus and Mountjoy (1969) also claimed that their findings were in line with earlier aerospace research, reviewed by Baker (1965), which indicated a vestibular effect on visual acuity, brightness discrimination, and visual illusions. However, close reading shows that the only vestibular effects that had been well established were those relating to movement illusions.

12 Whiteside and Campbell (1959) suggested that the observers in the centrifuge may have been subject to a reflex compensatory backward movement of the head, which might account for the increase in perceived size in accordance with Emmert's law.

13 It is also unclear what degree of natural head tilt the tumbling phase of the centrifuge might represent in terms of physics, or whether the calculated effect would be equivalent to the physiological effect (Wade and Day 1967) or the perceptual effect (Witkin 1964).

14 The evidence from reflex eye movements suggests that the simulation is successful (Jonkees and Philipszoon 1962).

15 The swing had a half amplitude of 38 cm and a swing period of 3 seconds, which gave a peak acceleration of 0.17 g, equivalent to a head tilt of 9.8 degrees.

16 In another experiment with 12 observers, Gregory and Ross (1964b) found similar directional differences with binocular viewing. They suggested that greater experience of forwards motion enables observers to interpret vestibular information for that direction more efficiently.

17 Van Eyl (1972) used a parallel swing with a peak acceleration of 0.13 g, equivalent to about 7.5 deg of head tilt.

18 A further difficulty is that the display time for the comparison stimulus was very brief (200 ms). This is hardly long enough for the observer to fixate the target in central vision, let alone to make a judgement of comparative size.

19 Wolbarsht and Lockhead (1985) and Lockhead and Wolbarsht (1989, 1991) called the reduction in perceived size the 'toy illusion', and related it to empty-field myopia.

20 A personal communication reported in Ross (1982).

21 Dember and Warm (1979).

22 The *Hunstville Times*, 28 January 1981.

23 The *Durham Sun* (29 June 1982, pp. 7–8), cited by Lockhead and Wolbarsht (1989).

24 BBC1 television programme, *For All Mankind*, shown 20 July 1989.

25 Hamilton (1964); Schmidt (1964).

26 The reports of Lichtenberg and Ockels are by personal communication.

27 Reported in Ross (1982).

28 Reported in Baird (1982).

29 This apparent movement is known as the *oculoagravic* or *elevator* illusion. See Benson (1965), Howard and Templeton (1966), pp. 425–6.

30 Ross (1982) also attempted to view an afterimage with open eyes against a visual background during the parabola. This proved difficult because of reflex eye movements and the constantly changing scene. The afterimage was scarcely visible against a bright sky during the zero-gravity phase. During the high-g phases it tended to float down over the earth's surface, and to appear smaller as it descended to the nearer foreground. Perceived size seemed to be related to the visual background, in accordance with Emmert's law. However, a vestibular contribution cannot be ruled out. It may be that attempts to fixate an object (such as the sun) prevent eye movements and eliminate changes in the visual background, thus destroying effects that occur with a free-floating image.

31 Under constant zero gravity the semicircular canals operate normally, but head tilt does not stimulate the otoliths. Head movements therefore produce an unusual combination of signals, causing nausea and illusory movement sensations.

32 It is interesting to note that Békésy (1949) reported the existence of a 'kinaesthetic analogue' of the moon illusion. The separation of two wires, as judged by touch by a blindfold observer, appeared to decrease as the wires were moved above or below the head. This remained true when the observer lay on his back, so the findings related to arm position rather than gravity. Békésy offered no explanation – he thought the moon illusion was due to visual size constancy and perceptual learning, and was not related to auditory or tactile analogues.

33 Plotinus. *Second Ennead*, VIII, trans. MacKenna and Page (1952), pp. 64–5.

34 See Morgan (1977) for a wide-ranging discussion of the philosophical and psychological background.

35 See reviews by Welch (1978, 1986).

36 See the review by Marks (1978) in his encyclopaedic book on the unity of the senses.

37 See review by Milner (1997).

38 Some examples of dorsal stream activities are: babies are able to set the gap between finger and thumb at the correct size to grasp an object within their reach (von Hofsten and Ronnqvist 1988); grip size is set by adults well in advance of contact with the target (Jakobson and Goodale 1991; Jeannerod 1981); and adults are able to judge which apertures are large enough to walk through (Warren and Whang 1987).

39 McKee and Welch (1992) showed that discrimination was better for linear size than for angular size at close viewing distances, and argued that linear size was the primary percept.

40 For example, Aglioti *et al.* (1995); Brenner and Smeets (1996).

41 For example, Daprati and Gentilucci (1997); Franz *et al.* (1998).

42 Bingham *et al.* (2000); DeLucia *et al.* (2000).

43 Observations and experiments have given mixed results. It has been claimed by Bourdon (1902), Taylor (1941), and others that the perceived size of afterimages viewed in the dark varies with imagined differences in projection distance, or when the observer moves his head towards the perceived location of the image. Taylor claimed that it only occurred if the eyes also converged. Gregory, Wallace, and Campbell (1959) stated that an afterimage projected onto the hand sometimes appeared to expand when the hand was brought nearer the body – a finding that contradicts Emmert's law. Davies (1973) had eight observers view an afterimage of a corridor while walking backwards in the dark: four reported that the end wall moved back with them whilst appearing to increase in size 'like walking along a corridor that was expanding'; two said that the end wall remained the same size and appeared to recede; and two reported no changes. Carey and Allan (1996) found that an afterimage of the hand appeared to increase with the distance of that hand, the extent being the same for both active and passive movement. Mon-Williams *et al.* (1997) found that vergence eye movements were both necessary and sufficient to cause the changes in perceived size. Bross (2000) found that the positive after-image of an object held in the hand did obey Emmert's law, but the afterimage of the hand itself did not shrink sufficiently at close distances. The topic thus remains controversial, with varied effects being reported.

44 Brenner, van Damme, and Smeets (1997) argued that kinaesthetic information about an object's distance did not improve visual judgements of its size.

45 It is possible that the binocular stereoscopic experiment of Kaufman and Kaufman (2000), which was outlined in Chapter 9, might constitute such evidence.

Chapter 13

1 This has been attempted by Higashiyama and Shimono (1994) who made use of islands and other landmarks at distances of up to 4.8 km.

2 Even Boring in his historical writings (e.g. 1942, pp. 223, 298) was guilty of this error.

3 Several recent authors have taken that approach (e.g. Day and Parks 1989; Haber and Levin 1989, 2001; Meehan and Day 1995). The similarities and differences between Berkeley and Gibson on this issue are discussed by Schwartz (1994) – a task he found 'more daunting in the case of Gibson than of Berkeley', because Gibson's shifting positions have been interpreted in conflicting ways by his supporters. See also Ross (2002).

4 The evidence from certain size-constancy experiments suggests that at short viewing distances linear size is the dominant percept (McKee and Smallman 1998). A different conclusion was reached by Kaneko and Uchikawa (1997), who argued that judgements of angular and linear size are affected by different cues, and could therefore be processed independently.

5 For example, Franks (1994).

6 Murray (1988), p. 435.

7 The most recent edition of Goldstein's perception textbook (1999, pp. 256–8) includes angular size and multi-component accounts of the illusion.

References

Abbott, T. K. (1864). *Sight and touch.* London: Longman, Green, Longman, Roberts and Green.

Aglioti, S., Goodale, M. A., and DeSouza, J. F. X. (1995). Size–contrast illusions deceive the eye but not the hand. *Current Biology*, **5**, 679–85.

Alexander, K. R. (1975). On the nature of accommodative micropsia. *American Journal of Optometry and Physiological Optics*, **52**, 79–84.

Allander, A. (1901). La grandeur apparente du soleil et de la lune. *Bulletin de la Société astronomique de France*, March, 139–41.

Allen, C. W. (1973). *Astrophysical quantities* (3rd edn). London: Athlone Press.

Allesch, G. J. von (1931). Zur nichteuklidischen Struktur des phänomenalen Raumes (Versuche an Lemur mongoz mongoz L.). Jena: Gustav Fischer.

Ames, A. (1946). Some demonstrations concerned with the origin and nature of our sensations. A laboratory manual. Dartmouth, NH: Dartmouth Eye Institute.

Andrews, D. P. (1964). Error-correcting perceptual mechanisms. *Quarterly Journal of Experimental Psychology*, **16**, 102–15.

Angell, F. (1924). Notes on the horizon illusion: I. *American Journal of Psychology*, **35**, 98–102.

Angell, F. (1932). Notes on the horizon illusion: II. *Journal of General Psychology*, **6**, 133–56.

Ariotti, P. E. (1973a). On the apparent size of projected after-images: Emmert's or Castelli's law? A case of 242 years anticipation. *Journal of the History of the Behavioral Sciences*, **9**, 18–28.

Ariotti, P. E. (1973b). A little known early 17th century treatise on vision: Benedetto Castelli's Discorso sopra la vista (1639, 1669). *Annals of Science*, **30**, 1–30.

Ariotti, P. E. (1973c). Benedetto Castelli and George Berkeley as anticipators of recent findings on the moon illusion. *Journal of the History of the Behavioral Sciences*, **9**, 328–32.

Aristotle (384–322 BC). *The works of Aristotle translated into English* (ed. J. A. Smith and W. D. Ross), Vol. 3: *Meteorologica* (trans. E. W. Webster); *De sensu* and *De somniis* form part of the *Parva naturalia* (trans. J. I. Beare). Oxford: Clarendon Press, 1908–52.

Aristotle (384–322 BC). *Meteorologica* (trans. H. D. P. Lee) Loeb Classical Library, London: Heinemann, 1962.

Ashley, M. L. (1898). Concerning the significance of intensity of light in visual estimates of distance. *Psychological Review*, **5**, 595–615.

Atkinson, K. (1997). *Human croquet.* London: Doubleday.

Backhouse, T. W. (1891). Apparent size of objects near the horizon. *Nature*, **45**, 7–8.

Bacon, R. (*c.* 1263). *Perspectiva.* In J. H. Bridges (ed.), *The 'Opus majus' of Roger Bacon*, Vol. 2. Oxford: Clarendon Press, 1897–1900.

Bailey, C. (ed.) (1926). *Epicurus. The extant remains.* Oxford: Clarendon Press.

Bailey, C. (ed.) (1947). *Lucretius: De rerum natura*, 3 vols. Oxford: Clarendon Press.

Baird, J. C. (1970). *Psychophysical analysis of visual space.* Oxford: Pergamon Press.

Baird, J. C. (1982). The moon illusion: II. A reference theory. *Journal of Experimental Psychology*, **111**, 304–15.

Baird, J. C. and Wagner, M. (1982). The moon illusion: I. How high is the sky? *Journal of Experimental Psychology*, **111**, 296–303.

Baird, J. C., Gulick, W. L., and Smith, W. M. (1962). The effects of angle of regard on the size of afterimages. *Psychological Record*, **12**, 263–71.

Baird, J. C., Wagner, M., and Fuld, K. (1990). A simple but powerful theory of the moon illusion. *Journal of Experimental Psychology: Human Perception and Performance*, **16**, 675–7.

Baker, G. A. (1965). *Visual capabilities in the space environment.* London: Pergamon.

Balfour, R. (1605). *Cleomedis meteora Graece et Latine.* Bordeaux: S. Milangius.

Barlow, H. B. and Mollon, J. D. (ed.) (1982). *The senses.* Cambridge: Cambridge University Press.

Bauer, H. (1911). *Die Psychologie Alhazens auf Grund von Alhazens Optik* (Beiträge zur Geschichte der Philosophie des Mittelalters, Band 10, Heft 5). Münster: Aschendorff.

Békésy, G. von (1949). The moon illusion and similar auditory phenomena. *American Journal of Psychology*, **62**, 540–52.

Benel, R. A. (1979). *Visual accommodation, the Mandelbaum effect, and apparent size.* Technical Report BEL-79–1/AFOSR-79–5, Las Cruces, New Mexico: New Mexico State University, Behavioral Engineering Laboratory.

Benson, A. J. (1965). Spatial disorientation in flight. In J. A. Gillies (ed.), *A textbook of aviation physiology*, pp. 1086–129. Oxford: Pergamon.

Berkeley, G. (1709). *An essay towards a new theory of vision.* In A. A. Luce and T. E. Jessop (ed.), *The works of George Berkeley bishop of Cloyne*, Vol. 1, pp. 143–239. London: T. Nelson and Sons, 1948.

Bettini, M. (1642). *Apiaria universae philosophiae mathematicae.* Bononiae: I. B. Ferronij.

Bevan, W. and Dukes, W. F. (1953). Color as a variable in the judgement of size. *American Journal of Psychology*, **66**, 283–8.

Bilderback, L. G., Taylor, R. E., and Thor, D. H. (1964). Distance perception in darkness. *Science*, **145**, 294–5.

Bingham, G. P. (1993). Perceiving the size of trees: biological form and the horizon ratio. *Perception and Psychophysics*, **54**, 485–95.

Bingham, G. P., Zaal, F., Robin, D., and Shull, J. A. (2000). Distortions in definite distance and shape perception as measured by reaching without and with haptic feed-back. *Journal of Experimental Psychology: Human Perception and Performance*, **26**, 1436–60.

Biot, J. B. (1810). *Traité élémentaire d'astronomie physique, etc* (2nd edn). Paris: Kloosterman.

Birch, T. (1756–7). *The history of the Royal Society of London for improving of natural knowledge, from its first rise*. London: A. Millar. Facsimile published as no. 44 in the *Sources of Science* series; New York: Johnson Reprint Corporation, 1968, Vol. 3, pp. 502–3.

Blondel, M. (1888). L'agrandissement des astres à l'horizon. *Revue Philosophique de la France et de l'etranger*, **27**, 197–9.

Blumenthal, H. J. (1971). *Plotinus' psychology: his doctrines of the embodied soul*. The Hague: Martinus Nijhoff.

Boff, K. R. and Lincoln, J. E. (1988). *Engineering data compendium: human perception and performance*. Ohio: AAMRL, Wright-Patterson Air Force Base.

Boring, E. G. (1940). Size constancy and Emmert's law. *American Journal of Psychology*, **53**, 293–5.

Boring, E. G. (1942). *Sensation and perception in the history of experimental psychology*. New York: Appleton-Century-Crofts.

Boring, E. G. (1943). The moon illusion. *American Journal of Physics*, **11**, 55–60.

Boring, E. G. (1948). The nature of psychology. In E. G. Boring, H. S. Langfeld, and H. P. Weld (ed.), *Foundations of psychology*, pp. 1–18. New York: Wiley.

Boring, E. G. (1962). [On the moon illusion]. *Science*, **137**, 902–6.

Boring, E. G. and Holway, A. H. (1940). Perceived size of the moon as a function of angle of regard. *Science*, **91**, 479–80.

Bourdon, B. (1897). Les résultats des travaux récents sur la perception visuelle de la profondeur. *Année Psychologique*, **4**, 390–431.

Bourdon, B. (1898). [Grandeur apparente de la lune]. *L'Intermédiaire des Biologistes*, **1**, 392–4.

Bourdon, B. (1899). Les objets paraissent-ils se rapetisser en s'élevant au-dessus de l'horizon? *Année Psychologique*, **5**, 55–64.

Bourdon, B. (1902). *La perception visuelle de l'espace*. Paris: Schleicher Frères.

Bowen, A. C. and Goldstein, B. R. (1996). Geminus and the concept of mean motion in Greco-Latin astronomy. *Archive for History of Exact Sciences*, **50**, 157–85.

Bowen, A. C. and Todd, R. B. (in press). *Physics and astronomy in later Stoic philosophy: Cleomedes' Meteora ('The Heavens')*. Berkeley: University of California Press.

Brenner, E. and Smeets, J. B. J.(1996). Size illusion influences how we lift but not how we grasp an object. *Experimental Brain Research*, **111**, 473–6.

Brenner, E., van Damme, W. J. M., and Smeets, J. B. J. (1997). Holding an object one is looking at: kinaesthetic information on the object's distance does not improve visual judgments of its size. *Perception and Psychophysics*, **59**, 1153–9.

Brentano, F. (1892). Ueber ein optisches Paradoxen. *Zeitschrift für Psychologie*, **3**, 349–58.

Brewer, Rev. Dr (1863). *A guide to the scientific knowledge of things familiar* (18th edn). London: Jarrold and Sons.

Brewster, D. (1849). Account of a new stereoscope. In N. J. Wade (ed.), *Brewster and Wheatstone on vision*, pp. 135–7. London: Academic Press, 1983.

Bross, M. (2000). Emmert's law in the dark: active and passive proprioceptive effects on positive visual afterimages. *Perception*, **29**, 1385–91.

Brunswik, E. (1944). Distal focussing of perception: size constancy in a representative sample of situations. *Psychological Monographs*, **254** (whole number).

Brunswik, E. (1956). *Perception and the representative design of psychological experiments.* Berkeley: University of California Press.

Burnet, J. (1773). *Of the origin and progress of language*, Vol. 1. Edinburgh: Kincaid and Creech, p. 31. Reprinted Menston: Scolar Press, 1967.

Burton, H. E. (1945). The Optics of Euclid. *Journal of the Optical Society of America*, **35**, 357–72.

Cambridge Anthropological Expedition to the Torres Straits (1901). *Reports*, Vol. 2, Part 1. Cambridge: Cambridge University Press.

Campbell, F. W. (1954). The minimum quantity of light required to elicit the accommodation reflex. *Journal of Physiology*, **123**, 357–66.

Carey, D. P. and Allan, K. (1996). *A motor signal and 'visual' size perception. Experimental Brain Research*, **110**, 482–6.

Carlson, V. R. (1977). Instructions and perceptual constancy judgements. In W. Epstein (ed.), *Stability and constancy in visual perception: mechanisms and processes*, pp. 217–54. New York: Wiley.

Carter, D. S. (1977). The moon illusion: a test of the vestibular hypothesis under monocular viewing conditions. *Perceptual and Motor Skills*, **45**, 1127–30.

Castelli, B. (1639). *Discorso sopra la vista.* In *Alcuni opuscoli filosofici del Padre D. Benedetto Castelli.* Bologna: G. Monti, 1669. Translated in Ariotti (1973b).

Cave, C. J. P. (1938). Apparent enlargement of the sun at the time of rising and setting. *Nature*, **141**, 290.

Ching, C., Peng, J., and Fang, Y. (1963a). The effect of distance and posture of observer on the perception of size. *Acta Psychologica Sinica*, **22**, 20–30.

Ching, C., Peng, J., Fang, Y., and Lin, C. (1963b). Size judgements of an object in elevation and in descent. *Acta Psychologica Sinica*, **24**, 175–85.

Church, J. (1960). Perceptual constancy. *Science*, **131**, 338.

Claparède, E. (1906a). L'agrandissement et la proximité apparente de la lune à l'horizon. *Archives de Psychologie*, **5**, 121–48.

Claparède, E. (1906b). A propos de la grandeur de la lune à l'horizon. *Archives de Psychologie*, **5**, 254–7.

Clausius, R. J. E. (1850). Uebersichtliche Darstellung der in das Gebiet der meteorologischen Optik gehörenden Erscheinungen. In J. A. Grunert (ed.), *Beiträge zur meteorologischen Optik und zuverwandten Wissenschaften, etc.*, Vol. 4, p. 369. Leipzig: E. B. Schwickert.

Cleomedes (*c.* AD third century). *Meteora* or *De motu circulari corporum caelestium*. See Goulet (1980) and Todd (1990).

Cohen, L. A. (1960). The moon illusion. *Science*, **131**, 694.

Cohen, M. R. and Drabkin, I. E. (1948). *A source book in Greek science.* New York: McGraw-Hill.

Conrad, J. (2000). *Getting the right exposure when photographing the moon.* http://www.calphoto.com/moon.htm. Downloaded 18 February 2001.

Cope, P. (1975). The moon illusion. *Journal of the British Astronomical Association*, **86**, 44–6.

Coren, S. (1989). The many moon illusions: an integration through analysis. In M. Hershenson (ed.), *The moon illusion*, pp. 351–70. Hillsdale, NJ: Erlbaum.

Coren, S. (1992). The moon illusion: a different view through the legs. *Perceptual and Motor Skills*, **75**, 827–31.

Coren, S. and Aks, D. J. (1990). Moon illusion in pictures: a multimechanism approach. *Journal of Experimental Psychology: Human Perception and Performance*, **16**, 365–80.

Coren, S. and Girgus, J. S. (1978). *Seeing is deceiving: the psychology of visual illusions.* Hillsdale, NJ: Erlbaum.

Coren, S., Ward, L. M., and Enns, J. T. (1994). *Sensation and perception* (4th edn). Fort Worth: Harcourt Brace College Publishers.

Cornish, V. (1935). *Scenery and the sense of sight.* Cambridge: Cambridge University Press.

Cornish, V. (1937). Apparent enlargement of the sun at the time of rising and setting. *Nature*, **140**, 1082–3.

Craddick, R. A. (1963). Height of Christmas tree drawings as a function of time. *Perceptual and Motor Skills*, **17**, 335–9.

Cutting, J. E. (1986). *Perception with an eye for motion.* Cambridge, MA: MIT Press.

Dadourian, H. M. (1946). The moon illusion. *American Journal of Physics*, **14**, 65–6.

Daprati, E. and Gentilucci, M. (1997). Grasping an illusion *Neuropsychologia*, **35**, 1577–82.

Da Silva, J. A. (1989). Gogel's laws and the simulated moon illusion in a large open field. In M. Hershenson (ed.), *The moon illusion*, pp. 319–41. Hillsdale, NJ: Erlbaum.

da Vinci, Leonardo (1452–1519). *The literary works of Leonardo da Vinci* (2nd edn; compiled and ed. J. P. Richter). Oxford: Oxford University Press, 1939.

da Vinci, Leonardo (1452–1519). *The literary works of Leonardo da Vinci: commentary*, by C. Pedretti. Oxford: Phaidon Press, 1977.

da Vinci, Leonardo (1452–1519). *The notebooks of Leonardo da Vinci* (trans. Edward MacCurdy). London: Jonathan Cape, 1956.

Davies, P. (1973). Effects of movements upon the appearance and duration of a prolonged visual afterimage: 2. Changes arising from movement of the observer in relation to the previously afterimaged scene. *Perception*, **2**, 155–60.

Day, R. H. and Parks, T. E. (1989). To exorcise a ghost from the perceptual machine. In M. Hershenson (ed.), *The moon illusion*, pp. 343–50. Hillsdale, NJ: Erlbaum.

Day, R. H., Stuart, G. W., and Dickinson, R. G. (1980). Size constancy does not fail below half a degree. *Perception and Psychophysics*, **28**, 263–5.

Dees, J. W. (1966a). Accuracy of absolute visual distance and size estimation in a space as a function of stereopsis and motion parallax. *Journal of Experimental Psychology*, **72**, 466–76.

Dees, J. W. (1966b). Moon illusion and size–distance invariance: an explanation based upon an experimental artifact. *Perceptual and Motor Skills*, **23**, 629–30.

DeLucia, P. R., Tresilian, J. R., and Meyer, L. E. (2000). Geometrical illusions can affect time-to-contact estimation and mimed prehension. *Journal of Experimental Psychology: Human Perception and Performance*, **26**, 552–67.

Dember, H. and Uibe, M. (1918). Ueber die scheinbare Gestalt des Himmelsgewölbes. *Annalen der Physik*, 4th series, **55**, 387–96.

Dember, H. and Uibe, M. (1920). Versuch einer physikalischen Lösung des Problems der sichtbaren Grössenänderung von Sonne und Mond in verschiedenen Höhen über dem Horizont. *Annalen der Physik*, 4th series, **61**, 353–78.

Dember, W. N. and Warm, J. S. (1979). *Psychology of perception* (2nd edn). New York: Holt, Rinehart and Winston.

Desaguiliers, J. T. (1736a). An attempt to explain the phaenomenon of the horizontal moon appearing bigger than when elevated many degrees above the horizon: supported by an experiment. Communicated Jan. 30, 1734–5. *Philosophical Transactions of the Royal Society of London*, **39**, 390–2.

Desaguiliers, J. T. (1736b). An explication of the experiment made in May 1735, as a farther confirmation of what was said in a paper given in January 30, 1734–5. to account for the appearance of the horizontal moon seeming larger than when higher. *Philosophical Transactions of the Royal Society of London,* **39**, 392–4.

Descartes, R. (1637a). *Dioptrics.* In E. Anscombe and P. T. Geach (ed.), *Descartes: philosophical writings.* London: Nelson's University Paperbacks/Open University, 1970.

Descartes, R. (1637b). *Discourse on method, optics, geometry, and meteorology* (trans. P. J. Olscamp). Indianapolis: Bobbs-Merrill, 1965.

Dicks, D. R. (1970). *Early Greek astronomy to Aristotle.* Ithaca, NY: Cornell University Press.

Dobbins, A. C., Jeo, R. M., Fiser, J., and Allman, J. M. (1998). Distance modulation of neural activity in the visual cortex. *Science,* **281**, 552–5.

Dunn, S. (1762). An attempt to assign the cause, why the sun and moon appear to the naked eye larger when they are near the horizon. *Philosophical Transactions of the Royal Society of London,* **52**, 462–73.

Duverney, G. J. (1683) *Traité de l'organe de l'ouie.* Paris: Estienne Michallet.

Edelstein, L. and Kidd, I. G. (ed.) (1972). *Posidonius. I. The fragments.* Cambridge: Cambridge University Press.

Edgerton, S. Y. (1975). *The Renaissance rediscovery of linear perspective.* New York: Basic Books.

Eginitis, D. (1898). Sur l'agrandissement des disques du soleil et de la lune à l'horizon. *Bulletin de la Société Astronomique de France,* October, 430–2.

Egusa, H. (1983). Effects of brightness, hue and saturation on perceived depth between adjacent regions in the visual field. *Perception,* **12**, 167–75.

Emmert, E. (1881). Grössenverhältnisse der Nachbilder. *Klinische Monatsblätter für Augenheilkunde und für augenärztliche Fortbildung,* **19**, 443–50.

Enright, J. T. (1975). The moon illusion examined from a new point of view. *Proceedings of the American Philosophical Society,* **119**, 87–117.

Enright, J. T. (1989a). The eye, the brain, and the size of the moon: toward a unified oculomotor hypothesis for the moon illusion. In M. Hershenson (ed.), *The moon illusion,* pp. 59–122. Hillsdale, NJ: Erlbaum.

Enright, J. T. (1989b). Manipulating stereopsis and vergence in an outdoor setting: moon, sky and horizon. *Vision Research,* **29**, 1815–24.

Epicurus (341–270 BC). Letter to Pythocles. In C. Bailey, *Epicurus: the extant remains,* pp. 56–81. Oxford: Clarendon Press, 1926.

Epstein, W. (1977). Historical introduction to the constancies. In W. Epstein (ed.), *Stability and constancy in visual perception,* pp. 1–22. New York: Wiley.

Epstein, W., Park, J., and Casey, A. (1961). The current status of the size–distance hypotheses. *Psychological Bulletin,* **58**, 491–514.

Euler, L. (1762). *Letters of Euler to a German princess, on different subjects in physics and philosophy,* Vol. 2: *Letters of April 20 to May 4, 1762* (trans. Henry Hunter). London: H. Murray, 1795.

Explanatory supplement to the astronomical ephemeris and the American ephemeris and nautical almanac (1961). London: Her Majesty's Stationery Office.

Farnè, M. (1977). Brightness as an indicator to distance: relative brightness per se or contrast with the background? *Perception*, **6**, 287–93.

Fechner, G. T. (1860). *Elemente der Psychophysik*. Leipzig: Breitkopf and Härtel.

Ferguson, K. (1999). *Measuring the universe: the historical quest to quantify space*. London: Headline.

Filehne, W. (1894). Die Form des Himmelsgewölbes. *(Pflügers) Archiv für Physiologie*, **59**, 279–308.

Filehne, W. (1910a). Ueber die Rolle des Erfahrungsmotive beim einaugigen perspektivischen Fernsehen. *Archiv für Anatomie und Physiologie, Physiologische abteilung*, 392–400.

Filehne, W. (1910b). Ueber die Betrachtung des Gestirne mittels Rauchgläser und über die verkleinernde Wirkung der Blickerhebung. *Archiv für Anatomie und Physiologie, Physiologische abteilung*, 523–30.

Filehne, W. (1912). Ueber die scheinbare Form des Himmelsgewölbes und die scheinbare Grösse der gestirne und Sternbilder. *Deutsche Revue*, November/December. Abstract in *Zeitschrift für Psychologie* (1920), **85**, 346–7.

Filehne, W. (1917). Der absolute Grösseneindruck beim sehen der irdischen Gegenstände und der Gestirne. *Archiv für Anatomie und Physiologie, Physiologische abteilung*, **41**, 197–221.

Filehne, W. (1923). Ueber die scheinbare Gestalt des Himmelsgewölbes. *Zeitschrift für Sinnesphysiologie*, **54**, 1–8.

Fletcher, B. (1963). *A history of architecture on the comparative method*. London: Athlone Press.

Forrest, D. W. (1973). Experiments on the moon illusion. *Astronomy and Space*, **3**, 271–4.

Frankel, H. R. (1978). The importance of Galileo's nontelescopic observations concerning the size of the fixed stars. *Isis*, **69**, 77–82.

Franks, A. (1994). Ever wonder why? *Stirling News*, 28 July, p. 4.

Franz, V. (1919). [Untitled contribution.] *Prometheus*, **30**, 142–3.

Franz, V., Fahle, M., Gegenfurtner, K. R., and Bülthoff, H. H. (1998). Size–contrast illusions deceive grasping as well as perception. *Perception*, **27** (supplement), 140.

Fröbes, J. S. J. (1916). *Lehrbuch der experimentellen Psychologie*. Revised edn Freiburg im Bresgau: Herder, 1923.

Fry, G. A. (1952). Gilinsky's equations for perceived size and distance. *Psychological Review*, **59**, 244–5.

Fry, G. A. (1983). Basic concepts underlying graphical analysis. In C. M. Schor and K. J. Ciuffreda (ed.), *Vergence eye movements: basic and clinical aspects*, pp. 403–64. London: Butterworths.

Fry, G. A., Bridgman, C. S., and Ellerbrock, V. J. (1949). The effects of atmospheric scattering on binocular depth perception. *American Journal of Optometry*, **26**, 9–15.

Furlong, E. J. (1972). The varying visual sizes of the moon. *Astronomy and Space*, **2**, 215–17.

Gassendi, P. (1636–42). Epistolae quatuor de apparente magnetudine solis humilis et sublimis. In P. Gassendi, *Opera omnia*. Stuttgart: Friedrich Fromman, 1964.

Gauss, K. F. (1880). *Briefwechsel zwischen Gauss und Bessel*. Leipzig: W. Engelmann.

Gibson, J. J. (1950). *The perception of the visual world*. Boston: Houghton Mifflin.

Gibson, J. J. (1966). *The senses considered as perceptual systems*. Boston: Houghton Mifflin.

Gibson, J. J. (1979). *The ecological approach to visual perception*. Boston: Houghton Mifflin.

Gigerenzer, G. and Murray, D. J. (1987). *Cognition as intuitive statistics*. Hillsdale, NJ: Erlbaum.

Giles, H. A. (1923). *A history of Chinese literature*. New York: Appleton.

Gillam, B. (1998). Illusions at century's end. In J. Hochberg (ed.) *Handbook of perception and cognition* (2nd edn), pp. 95–136. London: Academic Press.

Gilinsky, A. S. (1951). Perceived size and distance in visual space. *Psychological Review*, **58**, 460–82.

Gilinsky, A. S. (1955). The effect of attitude upon the perception of size. *American Journal of Psychology*, **68**, 173–92.

Gilinsky, A. S. (1971). Comment: Adaptation level, contrast, and the moon illusion. In M. H. Appley (ed.), *Adaptation level theory*, pp. 71–9. New York: Academic Press.

Gilinsky, A. S. (1980). The paradoxical moon illusions. *Perceptual and Motor Skills*, **50**, 271–83.

Gilinsky, A. S. (1989). The moon illusion in a unified theory of visual space. In M. Hershenson (ed.), *The moon illusion*, pp. 167–92. Hillsdale, NJ: Erlbaum.

Goethe, J. W. (1810). *Zur farbenlehre*. In J. W. Goethe, *Gedenkausgabe der Werke, Briefe und Gespraeche*, Vol. 16 (2nd edn). Zurich: Artemis-Verlag, 1964.

Gogel, W. C. (1974). Cognitive factors in spatial response. *Psychologia*, **17**, 213–25.

Gogel, W. C. (1977). The metric of visual space. In W. Epstein (ed.), *Stability and constancy in visual perception*, pp. 129–81. New York: Wiley.

Gogel, W. C. and Mertz, D. L. (1989). The contribution of heuristic processes to the moon illusion. In M. Hershenson (ed.), *The moon illusion*, pp. 235–58. Hillsdale, NJ: Erlbaum.

Goldstein, A. G. (1959). Linear acceleration and apparent distance. *Perceptual and Motor Skills*, **9**, 267–9.

Goldstein, B. R. (1967). The Arabic version of Ptolemy's 'Planetary Hypotheses'. *Transactions of the American Philosophical Society*, new series, **57**, Part 4, 3–12.

Goldstein, E. B. (1999). *Sensation and perception* (5th edn). Pacific Grove: Brooks/Cole.

Goldstein, G. (1962). Moon illusion: an observation. *Science*, **138**, 1340–1.

Gordon, I. E. (1997). *Theories of visual perception* (2nd edn). New York: Wiley.

Goüye, T. (1700). Diverses observations de physique générale. I. *Histoire de l'Académie Royale des Sciences*, 8–10.

Granger, F. (ed. and trans.) (1970). *Vitruvius: On architecture*, 2 vols. Cambridge, MA: Harvard University Press.

Grant, E. (ed.) (1974). *A source book in medieval science.* Cambridge, MA: Harvard University Press.

Gray, L. H. (ed.) (1964). *The mythology of all races*, 12 vols. New York: Cooper Square.

Gregory, J. (1668). *Geometriae pars universalis.* Padua.

Gregory, R. L. (1963). Distortion of visual space as inappropriate constancy scaling. *Nature*, **203**, 1407.

Gregory, R. L. (1970). *The intelligent eye.* London: Weidenfeld and Nicolson.

Gregory, R. L. (1981). *Mind in science: a history of explanations in psychology and physics.* London: Weidenfeld and Nicolson.

Gregory, R. L. (1998). *Eye and brain* (5th edn). Oxford: Oxford University Press.

Gregory, R. L. and Ross, H. E. (1964a). Visual constancy during movement: 1. Effect of S's forward or backward movement on size constancy. *Perceptual and Motor Skills*, **18**, 3–8.

Gregory, R. L. and Ross, H. E. (1964b). Visual constancy during movement: 2. Size constancy using one or both eyes or proprioceptive information. *Perceptual and Motor Skills*, **18**, 23–6.

Gregory, R. L., Wallace, J. G., and Campbell, F. W. (1959). Changes in the size and shape of visual after-images observed in complete darkness during changes of position in space. *Quarterly Journal of Experimental Psychology*, **11**, 54–5.

Grijns, G. (1906). L'agrandissement apparent de la lune à l'horizon. *Archives de Psychologie*, **5**, 319–25.

Gruber, H. E., King, W. L., and Link, S. (1963). Moon illusion: an event in imaginary space. *Science*, **139**, 750–2.

Gubisch, R. W. (1966). Over-constancy and visual acuity. *Quarterly Journal of Experimental Psychology*, **18**, 366–8.

Guébhard, A. (1898). Grandeur apparente de la lune. *L'Intermédiaire des Biologistes*, **1**, 351–2.

Guthrie, W. K. C. (1962). *A history of Greek philosophy*, Vol. 1. Cambridge: Cambridge University Press.

Guttmann, A. (1903). Blickrichtung und Grössenschätzung. *Zeitschrift für Psychologie*, **32**, 333–45.

Haber, R. N. and Levin, C. A. (1989). The lunacy of moon watching: some preconditions on explanations of the moon illusion. In M. Hershenson (ed.), *The moon illusion*, pp. 299–317. Hillsdale, NJ: Erlbaum.

Haber, R. N. and Levin, C. A. (2001). The independence of size perception and distance perception. *Perception and Psychophysics*, **63**, 1140–52.

Haenel, H. (1909). Die Gestalt des Himmels und Vergrösserung der Gestirne am Horizonte. *Zeitschrift für Psychologie*, **51**, 161–99.

Hamilton, J. E. (1964). Factors affecting the moon illusion. *American Journal of Optometry and Archives of American Academy of Optometry*, **41**, 490–3.

Hamilton, J. E. (1965). Effect of observer elevation on the moon illusion. *Journal of Engineering Psychology*, **4**, 57–67.

Hamilton, J. E. (1966). Luminance and the moon illusion. *American Journal of Optometry and Archives of American Academy of Optometry*, **43**, 593–604.

Hamilton, W. (1849). *The works of Thomas Reid* (2nd edn). Edinburgh: Maclachlan, Stewart, and Co.

Hammer, L. R. (1962). *Perception of the visual vertical under reduced gravity* (AMRL-TDR-62–55). Ohio: Wright-Patterson Air Force Base.

Haskins, C. H. (1960). *Studies in the history of mediaeval science*. New York: Frederick Ungar.

Hatfield, G. C. and Epstein, W. (1979). The sensory core and the medieval foundations of early modern perceptual theory. *Isis*, **70**, 363–84.

Heelan, P. A. (1983). *Space-perception and the philosophy of science*. Berkeley: University of California Press.

Heinemann, E. G., Tulving, E., and Nachmias, J. (1959). The effect of oculomotor adjustments on apparent size. *American Journal of Psychology*, **72**, 32–45.

Helmholtz, H. von (1856–66). *Treatise on physiological optics* (ed. J. P. C. Southall; trans. of the 3rd German edn, 1909–11). New York: Dover, 1962.

Helson, H. (1964). *Adaptation level theory: an experimental and systematic approach to behavior*. New York: Harper.

Henning, H. (1919). Die besondere Funktionen der roten Strahlen bei der scheinbaren Grösse von Sonne und Mond am Horizont. *Zeitschrift für Sinnesphysiologie*, **50**, 275–310.

Hermans, T. G. (1954). The relationship of convergence and elevation changes to judgements of size. *Journal of Experimental Psychology*, **48**, 204–8.

Herschel, J. F. W. (1833). *A treatise on astronomy* (new edn). London: Longman, 1851.

Hershenson, M. (1982). Moon illusion and spiral aftereffect: illusions due to the loom-zoom system? *Journal of Psychology: General*, **111**, 423–40.

Hershenson, M. (1989a). Moon illusion as anomaly. In M. Hershenson (ed.), *The moon illusion*, pp. 123–46. Hillsdale, NJ: Erlbaum.

Hershenson, M. (1989b). The puzzle remains. In M. Hershenson (ed.), *The moon illusion*, pp. 377–384. Hillsdale, NJ: Erlbaum.

Heuer, H. and Owens, D. A. (1989). Vertical gaze direction and the resting posture of the eyes. *Perception*, **18**, 363–77.

Heuer, H., Wischmeyer, E., Brüwer, M., and Römer, T. (1991). Apparent size as a function of vertical gaze direction: new tests of an old hypothesis. *Journal of Experimental Psychology: Human Perception and Performance*, **17**, 232–45.

Higashiyama, A. (1992). Anisotropic perception of visual angle: implications for the horizontal–vertical illusion, over constancy of size, and the moon illusion. *Perception and Psychophysics*, **51**, 218–30.

Higashiyama, A. and Shimono, K. (1994). How accurate is size and distance perception for very far terrestrial objects? Function and causality. *Perception and Psychophysics*, **55**, 429–42.

Hirschberg, J. (1898). Die Optik der alten Griechen. *Zeitschrift für Psychologie*, **16**, 321–51.

Hobbes, T. (1658). *Physics or the phenomena of nature*. In T. Hobbes, *English Works*, Vol. 1 (trans. W. Molesworth). London: John Bohn, 1839.

Hofsten, C. von and Ronnqvist, L. (1988). Preparation for grasping an object: a developmental study. *Journal of Experimental Psychology: Human Performance and Perception*, **14**, 610–21.

Holland, G. (1958). Untersuchung über den Einfluss der Fixationsentfernung und Blickrichtung auf die horizontale Heterophorie (Exo- und Eso-Phorie). *Albrecht von Graefes Archiv für Ophthalmologie*, **160**, 144–60.

Holway, A. H. and Boring, E. G. (1940a). The moon illusion and the angle of regard. *American Journal of Psychology*, **53**, 109–16.

Holway, A. H. and Boring, E. G. (1940b). The apparent size of the moon as a function of the angle of regard: further experiments. *American Journal of Psychology*, **53**, 537–53.

Holway, A. H. and Boring, E G. (1940c). The dependence of apparent visual size upon illumination. *American Journal of Psychology*, **53**, 587–89.

Holway, A. H. and Boring, E. G. (1941). Determinants of apparent visual size with distance variant. *American Journal of Psychology*, **54**, 21–37.

Ho Peng Yoke (1966). *The astronomical chapters of the Chin Shu*. Paris: Mouton.

Howard, I. P. (1982). *Human visual orientation*. New York: Wiley.

Howard, I. P. and Templeton, W. B. (1966). *Human spatial orientation*. London: Wiley.

Howland, H. C. (1959). Moon illusion and age. *Science*, **130**, 1364–5.

Hull, J. C., Gill, R. T., and Roscoe, S. N. (1982). Locus of the stimulus to visual accommodation: where in the world, or where in the eye? *Human Factors*, **24**, 311–19.

Humboldt, A. von. (1850). *Views of nature.* (trans. E. C. Otté and H. G. Bohn). London: Henry, G. Bohn.

Humphreys, W. J. (1964). *Physics of the air.* New York: Dover.

Hutton, C. (1796). *Mathematical and philosophical dictionary.* London.

Huxley, T. H. (1885). *Lessons in elementary physiology* (revised edn). London: Macmillan.

Iavecchia, J. H., Iavecchia, H. P., and Roscoe, S. N. (1983). The moon illusion revisited. *Aviation, Space, and Environmental Medicine*, **54**, 39–46.

Ibn al-Haytham (*c.* 1040). *Optics.* See Sabra (1989).

Indow, T. (1968). Multidimensional mapping of visual space with real and simulated stars. *Perception and Psychophysics*, **3**, 45–53.

Ittelson, W. H. (1951). Size as a cue to distance: static localization. *American Journal of Psychology*, **64**, 54–67.

Jackson, K. H. (1971). *A Celtic miscellany: translations from the Celtic literatures.* Harmondsworth: Penguin.

Jakobson, L. S. and Goodale, M. A. (1991). Factors affecting higher-order movement planning: a kinematic analysis of human prehension. *Experimental Brain Research*, **86**, 199–208.

James, W. (1890). *The principles of psychology.* New York: Dover, 1950.

Jeannerod, M. (1981). Intersegmental coordination during reaching at natural objects. In J. Long and A. Baddeley (ed.), *Attention and performance*, Vol. IX, pp. 153–68. Hillsdale, NJ: Erlbaum.

Johannsen, D. E. (1971). Early history of perceptual illusions. *Journal of the History of the Behavioral Sciences*, **7**, 127–40.

John of Sacrobosco (d. 1250). *De sphaera.* In L. Thorndike, *The sphere of Sacrobosco and its commentators.* Chicago: University of Chicago Press, 1949.

Johns, E. H. and Sumner, F. C. (1948). Relation of the brightness difference of colors to their apparent distances. *Journal of Psychology*, **26**, 25–9.

Jonkees, L. B. W. and Philipszoon, H. J. (1962). Nystagmus provoked by linear acceleration. *Acta Physiologica et Pharmacologia Neerlandica*, **10**, 239.

Joynson, R. B. (1949). The problem of size and distance. *Quarterly Journal of Experimental Psychology*, **1**, 119–35.

Judd, C. H. (1897). Some facts of binocular vision. *Psychological Review*, **4**, 374–89.

Jung, R. (1961). Neuronal integration in the visual cortex and its significance for visual information. In W. A. Rosenblith (ed.), *Sensory communication*, pp. 627–74. New York: Wiley.

Kamman, R. (1967). The overestimation of vertical distance and slope and its role in the moon illusion. *Perception and Psychophysics*, **2**, 585–9.

Kanda, S. (1933). Ancient records of sunspots and auroras in the Far East and the variation of the period of solar activity. *Proceedings of the Imperial Academy of Japan*, **9**, 293–6.

Kaneko, H. and Uchikawa, K. (1997). Perceived angular and linear size: the role of binocular disparity and visual surround. *Perception*, **26**, 17–27.

Kaufman, L. (1960). An investigation of the moon illusion. Ph.D. dissertation, New School for Social Research, New York.

Kaufman, L. (1974). *Sight and mind: an introduction to visual perception.* New York: Oxford University Press.

Kaufman, L. (1979). *Perception: the world transformed.* New York: Oxford University Press.

Kaufman, L. and Kaufman, J. H. (2000). Explaining the moon illusion. *Proceedings of the National Academy of Sciences*, **97**, 500–5.

Kaufman, L. and Rock, I. (1962a). The moon illusion, I. *Science*, **136**, 953–61.

Kaufman, L. and Rock, I. (1962b). The moon illusion. *Scientific American*, **207** (1), 120–30.

Kaufman, L. and Rock, I. (1989). The moon illusion thirty years later. In M. Hershenson (ed.), *The moon illusion*, pp. 193–234. Hillsdale, NJ: Erlbaum.

Kemp, M. (1990). *The science of art. Optical themes in Western art from Brunelleschi to Seurat.* New Haven, CT: Yale University Press.

Kiesow, F. (1925/6). Ueber die Vergleichung linearer Strecken und ihre Bezeihung zum Weberschen Gesetze. *Archiv für die gesamte Psychologie* (1925), **52**, 61–90; (1925), **53**, 443–6; (1926), **56**, 421–51.

King, W. L. and Gruber, H. E. (1962). Moon illusion and Emmert's law. *Science*, **135**, 1125–6.

King, W. L. and Hayes, M. C. (1966). The sun illusion: individual differences in remembered size and distance judgements. *Psychonomic Science*, **5**, 65–6.

Kinsler, L. E. (1945). Imaging of underwater objects. *American Journal of Physics*, **13**, 255–7.

Kishto, B. N. (1965). The colour stereoscopic effect. *Vision Research*, **5**, 313–29.

Knoll, H. A. (1952). A brief history of 'nocturnal myopia' and related phenomena. *American Journal of Optometry and Physiological Optics*, **29**, 69–81.

Koffka, K. (1936). *The principles of gestalt psychology.* New York: Harcourt Brace.

Kolb, R. (1999). Blind watchers of the sky: the people and ideas that shaped our view of the universe. Oxford: Oxford University Press.

Lambercier, M. (1946). La constance des grandeurs en comparaisons serials. *Archives de Psychologie, Genève*, **31**, 79–282.

Laming, D. J. (1986). *Sensory analysis.* London: Academic Press.

Le Cat, C. N. (1744). *Traité des sens* (new edition). Rouen.

Le Conte, J. (1881). *Sight: an exposition of the principles of monocular and binocular vision*. International Scientific Series, Vol. 33. London: Kegan Paul.

Leibowitz, H. and Hartman, T. (1959a). Magnitude of the moon illusion as a function of the age of the observer. *Science*, **130**, 569–70.

Leibowitz, H. and Hartman, T. (1959b). Reply to Howland. *Science*, **130**, 1365–6.

Leibowitz, H. and Hartman, T. (1960a). Reply to Church. *Science*, **131**, 239.

Leibowitz, H. and Hartman, T. (1960b). Reply to Cohen. *Science*, **131**, 694.

Leibowitz, H. W. and Harvey, L. O. (1967). Size matching as a function of instructions in a naturalistic environment. *Journal of Experimental Psychology*, **74**, 378–82.

Leibowitz, H. W. and Harvey, L. O. (1969). Effect of instructions, environment, and type of test object on matched size. *Journal of Experimental Psychology*, **81**, 36–43.

Leibowitz, H. W. and Owens, D. A. (1975). Anomalous myopias and the intermediate dark focus of accommodation. *Science*, **189**, 646–8.

Leibowitz, H. W. and Owens, D. A. (1989). Multiple mechanisms of the moon illusion and size perception. In M. Hershenson (ed.), *The moon illusion*, pp. 281–6. Hillsdale, NJ: Erlbaum.

Leibowitz, H. W., Hennessy, R. T., and Owens, D. A. (1975). The intermediate resting position of accommodation and some implications for space perception. *Psychologia*, **18**, 162–70.

Leiri, F. (1931). Ueber die Bedeutung der roten Strahlen bei der scheinbaren Vergrösserung von Sonne und Mond am Horizont. *Zeitschrift für Sinnesphysiologie*, **61**, 325–34.

Lejeune, A. (ed. and trans.) (1989). *L'Optique de Claude Ptolémée* (2nd edn). Leiden: E. J. Brill.

Levene, J. R. (1965). Nevil Maskelyne, FRS, and the discovery of night myopia. *Royal Society of London, Notes and Records*, **20**, 100–8.

Levitt, I. M. (1952). Moon illusion. *Sky and Telescope*, **11**, 135–6.

Lewis, R. T. (1862). On the changes in the apparent size of the moon. *Philosophical Magazine*, **23**, 380–2.

Lewis, R. T. (1898). Grandeur apparente de la lune. *L'Intermédiaire des Biologistes*, **1**, 391–2.

Lichten, W. and Lurie, S. (1950). A new technique for the study of perceived size. *American Journal of Psychology*, **63**, 280–2.

Lindberg, D. C. (1967). Alhazen's theory of vision and its reception in the West. *Isis*, **58**, 321–41.

Lindberg, D. C. (1970). The theory of pinhole images in the 14th century. *Archive for History of Exact Sciences*, **6**, 299–325.

Lindberg, D. C. (1976). *Theories of vision from Al-Kindi to Kepler*. Chicago: University of Chicago Press.

Lloyd, G. E. R. (1982). Observational error in later Greek science. In J. Barnes, J. Brunschwig, M. Burnyeat, and M. Schofield (ed.), *Science and speculation: studies in Hellenistic theory and practice*, pp. 128–64. Cambridge: Cambridge University Press.

Lockhead, G. R. and Wolbarsht, M. L. (1989). The moon and other toys. In M. Hershenson (ed.), *The moon illusion*, pp. 259–66. Hillsdale, NJ: Erlbaum.

Lockhead, G. R. and Wolbarsht, M. L. (1991). Toying with the moon illusion. *Applied Optics*, **30**, 3504–7.

Loftus, G. R. (1985). Size illusion, distance illusion, and terrestrial passage: comment on Reed. *Journal of Experimental Psychology: General*, **114**, 119–21.

Logan, J. (1736). Some thoughts concerning the sun and moon, when near the horizon, appearing larger than when near the zenith. *Philosophical Transactions of the Royal Society of London*, **39**, 404–5.

Lohmann, W. (1908). Ueber dem Fragen die Grössererscheinungen von Monde und sternen am Horizont und die scheinbare Form des Himmelsgewölbes. *Zeitschrift für Psychologie*, **51**, 154.

Lohmann, W. (1920). Ueber dem Fragen nach dem Grössererscheinen von Sonne, Mond und Sternen am Horizont und der scheinbaren Form des Himmelsgewölbes. *Zeitschrift für Sinnesphysiologie*, **51**, 96–120.

Loiselle, A. (1898). (Grandeur apparente de la lune). *L'Intermédiaire des Biologistes*, **1**, 352.

Lombardo, T. J. (1987). *The reciprocity of perceiver and environment: the evolution of James J. Gibson's ecological psychology*. Hillsdale, NJ: Erlbaum.

Luckiesh, M. (1918). On 'retiring' and 'advancing' colors. *American Journal of Psychology*, **29**, 182–6.

Luckiesh, M. (1921). The apparent form of the sky-vault. *Journal of the Franklin Institute*, **191**, 259–63.

Luckiesh, M. (1922). *Visual illusions – their causes, characteristics and applications*. Reprinted New York: Dover, 1965.

Lucretius (*c*. 95–55 BC). *De rerum natura*, 3 vols. (ed. with prolegomena, critical apparatus, translation and commentary C. Bailey). Oxford: Clarendon Press, 1947.

Lühr, K. (1898). Die scheinbare Vergrösserung der Gestirne in der Nähe des Horizontes. *Mitteilungen der Vereinigung von Freunden der Astronomie*, **8**, 31–5.

Lynch, D. K. and Livingston, W. (1995). *Color and light in nature*. Cambridge: Cambridge University Press.

MacKenna, S. and Page, B. S. *see* Plotinus.

Macrobius (early 5th century). *The Saturnalia* (trans. P. V. Davies). New York: Columbia University Press, 1969.

Malebranche, N. (1675). *The search after truth* (trans. T. M. Lennon and P. J. Olscamp). Columbus, Ohio: Ohio State University Press, 1980.

Malebranche, N. (1693). Réponse du P. Malebranche à M. Regis. In N. Malebranche, *Oeuvres complètes*, Vol. 17–1. Paris: J. Vrin, 1960.

Malebranche, N. (1694). Pièces diverses concernant la polémique Malebranche–Regis. In N. Malebranche, *Oeuvres complètes*, Vol.17–1. Paris: J. Vrin, 1960.

Marks, L. E. (1978). *The unity of the senses: interrelations among the modalities.* New York: Academic Press.

Marshack, A. (1975). Exploring the mind of Ice Age Man. *National Geographic Magazine*, **147** (1), 64–89.

Maskelyne, N. (1789). An attempt to explain a difficulty in the theory of vision, depending on the different refrangibility of light. *Philosophical Transactions of the Royal Society of London*, **79**, 256–64.

Mayr, R. (1904). Die scheinbare Vergrösserung von Sonne, Mond und Sternbildern am Horizont. *(Pflügers) Archiv für Physiologie*, **101**, 349–422.

McCready, D. (1965). Size–distance perception and accommodation-convergence micropsia – a critique. *Vision Research*, **5**, 189–206.

McCready, D. (1985). On size, distance, and visual angle perception. *Perception and Psychophysics*, **37**, 323–34.

McCready, D. (1986). Moon illusions redescribed. *Perception and Psychophysics*, **39**, 64–72.

McCready, D. (1999). *The moon illusion explained.* Article placed on website http: //facstaff.uww.edu/mccreadd/ on 18 May1999.

McKee, S. P. and Smallman, H. S. (1998). Size and speed constancy. In V. Walsh and J. J. Kulikowski (ed.), *Visual constancies: why things look as they do*, pp. 373–408. Cambridge: Cambridge University Press.

McKee, S. P. and Welch, L. (1992). The precision of size constancy. *Vision Research*, **32**, 1447–60.

McNeil, R. (2001). A little moonshine over Callanish relieves the stresses of the guys and gulls. *The Scotsman* (Edinburgh), 12 May 2001.

McNulty, J. A. and St Claire-Smith, R. (1964). Terrain effects upon perceived distance. *Canadian Journal of Psychology*, **18**, 175–82.

Meehan, J. W. and Day, R. H. (1995). Visual accommodation as a cue for size. *Ergonomics*, **38**, 1239–49.

Meili, R. (1960). Ueberlegungen zur Mondtäuschung. *Psychologische Beiträge*, **5**, 154–66.

Mellerio, J. (1966). Ocular refraction at low illuminations. *Vision Research*, **6**, 217–37.

Meyering, T. C. (1989). *Historical roots of cognitive science: the rise of a cognitive theory of perception from antiquity.* Dordrecht: Kluwer.

Middleton, W. E. K. (1958). *Vision through the atmosphere.* Toronto: University of Toronto Press.

Middleton, W. E. K. (1960). Bouguer, Lambert, and the theory of horizontal visibility. *Isis,* **51**, 145–9.

Miles, P. W. (1951). The relation of perceived size of half-images at the fusion level to projection on the horopter. *American Journal of Ophthalmology,* **34**, 1543–61.

Miller, A. (1943). Investigation of the apparent shape of the sky. B.Sc. thesis, Pennsylvania State College, University Park, PA 16802.

Milner, A. D. (1997). Vision without knowledge. *Philosophical Transactions of the Royal Society of London B,* **352**, 1249–56.

Minnaert, M. (1954). *The nature of light and colour in the open air* (trans. H. M. Kremmer-Priest, revised by K. E. Brian Jay). New York: Dover Publications.

Mollon, J. D. (1997). 'On the basis of velocity cues alone': some perceptual themes 1946–1996. *Quarterly Journal of Experimental Psychology,* **50A**, 859–78.

Mollon, J. D. and Ross, H. E. (1975). Gregory on the sun illusion. *Perception,* **4**, 115–18.

Molyneux, W. (1687). Concerning the apparent magnitude of the sun and moon, or the apparent distance of two stars, when nigh the horizon and when higher elevated. *Philosophical Transactions of the Royal Society of London,* **16**, 314–23.

Mon-Williams, M. and Tresilian, J. R. (1999). The size–distance paradox is a cognitive phenomenon. *Experimental Brain Research,* **126**, 578–82.

Mon-Williams, M., Tresilian, J. R., Plooy, A., Wann, J. P., and Broerse, J. (1997). Looking at the task in hand: vergence eye movements and perceived size. *Experimental Brain Research,* **117**, 501–6.

Morgan, M. J. (1977). *Molyneux's question: vision, touch and the philosophy of perception.* Cambridge: Cambridge University Press.

Morgan, M. J. (1992). On the scaling of size judgments by orientational cues. *Vision Research,* **32**, 1433–45.

Morinaga, S. (1935). Ueber die Blickrichtung und die Mondtäuschung (in Japanese). *Japanese Journal of Psychology,* **10**, 1–25. Johnson Reprint Corporation, 1966. English abstract in *Psychological Abstracts* (1935), **9**, No. 4940.

Müller, A. (1906). Le problème du grossissement apparent des astres à l'horizon considéré au point de vue méthodologique. *Archives de Psychologie,* **5**, 305–18.

Müller, A. (1907). Die referenzflächentheorie der Täuschung am Himmelsgewölbe und an den Gestirnen. *Zeitschrift für Psychologie,* **44**, 186–200.

Müller, A. (1918). *Die referenzflächen des Himmels und der Gestirne.* Braunschweig: Friedrich Vieweg und Sohn.

Müller, A. (1921). Beiträge zum Problem der Referenzflächen des Himmels und der Gestirne. *Archiv für die gesamte Psychologie,* **41**, 47–89.

Murray, D. J. (1988) *A history of Western psychology* (2nd edn). Englewood Cliffs: Prentice Hall.

Myers, C. S. (1911). *A textbook of experimental psychology* (2nd edn), Part 1 – Text-book. Cambridge: Cambridge University Press.

Needham, J. (1959). *Science and civilisation in China*, Vol. 3: *Mathematics and the sciences of the heavens and the earth.* Cambridge: Cambridge University Press.

Neugebauer, O. (1975). *A history of ancient mathematical astronomy.* New York: Springer-Verlag.

Oldfather, C. H. (trans.) (1967). *Diodorus of Sicily*, Vol. 2. Loeb Classical Library, London: Heinemann.

Omar, S. B. (1977). *Ibn al-Haytham's optics: a study of the origins of experimental science.* Minneapolis: Bibliotheca Islamica.

O'Neil, W. M. (1957). *An introduction to method in psychology.* Melbourne: Melbourne University Press.

O'Neil, W. M. (1969). Fact and theory: an aspect of the philosophy of science. Sydney: Sydney University Press.

Ono, H., Muter, P., and Mitson, L. (1974). Size–distance paradox with accommodative micropsia. *Perception and Psychophysics*, **15**, 301–7.

Oppel, J. J. (1855). Ueber geometrisch-optische Täuschungen. *Jahresberichte der physikalische verein, Frankfurt*, 37–47.

Orbach, J. and Solhkhah, N. (1968). Size judgements of discs presented against the zenith sky. *Perceptual and Motor Skills*, **26**, 371–4.

Osaka, R. (1962). Celestial illusion – an overview of the history and theories. *Psychologia*, **5**, 24–31.

O'Shea, R. P., Blackburn, S. G., and Ono, H. (1994). Contrast as a depth cue. *Vision Research*, **34**, 1595–1604.

Otero, J. M. (1951). Influence of the state of accommodation on the visual performance of the human eye. *Journal of the Optical Society of America*, **41**, 942–8.

Over, R. (1962a). Stimulus wavelength variation and size and distance judgements. *British Journal of Psychology*, **53**, 141–7.

Over, R. (1962b). Brightness judgements and stimulus size and distance. *British Journal of Psychology*, **53**, 431–8.

Owens, D. A. (1979). The Mandelbaum effect: evidence for an accommodative bias toward intermediate viewing distances. *Journal of the Optical Society of America*, **69**, 646–52.

Oyama T. and Yamamura, T. (1960). The effect of hue and brightness on the depth perception in normal and color-blind subjects. *Psychologia*, **3**, 191–4.

Pannekoek, A. (1961). *A history of astronomy.* New York: Interscience Publishers.

Parks, T. E. (1989). A brief comment. In M. Hershenson (ed.), *The moon illusion*, pp. 371–4. Hillsdale, NJ: Erlbaum.

Pastore, N. (1971). *Selective history of theories of visual perception: 1650–1950.* New York: Oxford University Press.

Payne, M. C. (1964). Color as an independent variable in perceptual research. *Psychological Bulletin*, **61**, 199–208.

Pecham, John (*c.* 1274). *Perspectiva communis.* In D. C. Lindberg, *John Pecham and the science of optics.* Madison: University of Wisconsin Press, 1970.

Pernter, J. M. and Exner, F. M. (1922). *Meteorologische Optik* (2nd edn). Vienna: W. Braumüller.

Piaget, J. and Lambercier, M. (1943). Recherches sur le développement des perceptions: III. Le problème de la comparaison visuelle en profondeur (constance de la grandeur) et l'erreur systématique de l'étalon. *Archives de Psychologie, Genève*, **29**, 255–308.

Pickford, R. W. (1943). Some effects of veiling glare in binocular vision. *British Journal of Psychology*, **33**, 150–61.

Pillsbury, W. B. and Schaefer, B. R. (1937). A note on 'advancing and retreating' colors. *American Journal of Psychology*, **49**, 126–30.

Plateau, J. (1839a). Note sur l'irradiation. *Bulletin de l'Académie Royale des Sciences, des Lettres et des Beaux-Arts de Belgique,* **6**, Part 1, 501–5.

Plateau, J. (1839b). Deuxième note sur l'irradiation. *Bulletin de l'Académie Royale des Sciences, des Lettres et des Beaux-Arts de Belgique,* **6**, Part 2, 102–6.

Plateau, J. (1880). Une application des images accidentelles. *Bulletin de l'Académie Royale de Belgique*, 2nd series, **49**, 316–19.

Plato (427–347 BC). *Timaeus* (trans. F. M. Cornford). Indianapolis: Bobbs-Merrill, 1959.

Plotinus (205–270). *The six Enneads* (trans. S. MacKenna and B. S. Page). In *Encyclopaedia Britannica great books of the western world*, Vol. 17. Chicago: William Benton, 1952.

Plug, C. (1989a). The registered distance of the celestial sphere: some historical cross-cultural data. *Perceptual and Motor Skills*, **68**, 211–17.

Plug, C. (1989b). Annotated bibliography. In M. Hershenson (ed.), *The moon illusion*, pp. 385–407. Hillsdale, NJ: Erlbaum.

Plug, C. and Ross, H. E. (1989). Historical review. In M. Hershenson (ed.), *The moon illusion*, pp. 5–27. Hillsdale, NJ: Erlbaum.

Plug, C. and Ross, H. E. (1994). The natural moon illusion: a multi-factor angular account. *Perception*, **23**, 321–33.

Ponzo, M. (1913). Rapports entre quelques illusions visuelles de contraste angulaire et l'appréciation de grandeur des astres à l'horizon. *Archives Italiennes de Biologie*, **58**, 327–9.

Porta, G. B. della (1593). *De refractione optices parte.* Naples: Carlinum and Pacum.

Porterfield, W. (1759). *A treatise on the eye, the manner and phaenomena of vision.* Edinburgh: A. Miller.

Poulton, E. C. (1989). *Bias in quantifying judgments.* Hove: Erlbaum.

Pozdena, R. F. (1909). Eine Methode zur experimentellen und konstruktiven Bestimmung der Form des Firmamentes. *Zeitschrift für Psychologie,* **51**, 200–46.

Ptolemaeus, Claudius. (*c.* 142). *The Almagest* (trans. R. C. Taliaferro). In *Encyclopaedia Britannica great books of the Western world,* Vol. 16. Chicago: William Benton, 1952.

Ptolemaeus, Claudius (*c.* 170). *L'optique de Claude Ptolémée* (ed. Albert Lejeune). Publications Universitaires de Louvain, 1956; 2nd edn Leiden: E. J. Brill, 1989.

Ptolemaeus, Claudius (*c.* 170). *Ptolemy's theory of visual perception: an English translation of the Optics* (trans. A. Mark Smith). *Transactions of the American Philosophical Society* (1996), **86**, Part 2.

Quantz, J. O. (1895). The influence of the color of surfaces on our estimation of their magnitude. *American Journal of Psychology,* **7**, 26–41.

Quiller-Couch, A. (ed.) (1955). *The Oxford book of ballads,* p. 329. Oxford: Clarendon Press.

Reed, C. F. (1984). Terrestrial passage theory of the moon illusion. *Journal of Experimental Psychology: General,* **113**, 489–500.

Reed, C. F. (1985). More things in heaven and earth: a reply to Loftus. *Journal of Experimental Psychology: General,* **114**, 122–4.

Reed, C. F. (1989). Terrestrial and celestial passage. In M. Hershenson (ed.), *The moon illusion,* pp. 267–78. Hillsdale, NJ: Erlbaum.

Reed, C. F. (1996). The immediacy of the moon illusion. *Perception,* **25**, 1295–1300.

Reed, C. F. and Krupinski, E. A. (1992). The target in the celestial (moon) illusion. *Journal of Experimental Psychology: Human Perception and Performance,* **18**, 247–56.

Regis, P. S. (1694). Première replique de M. Regis à la réponse du R. P. Malebranche … Summary in N. Malebranche, *Oeuvres complètes,* Vol.17–1. Paris: J. Vrin, 1960.

Reimann, E. (1902a). Die scheinbare Vergrösserung der Sonne und des Mondes am Horizont, I. Geschichte des Problems. *Zeitschrift für Psychologie,* **30**, 1–38.

Reimann, E. (1902b). Die scheinbare Vergrösserung der Sonne und des Mondes am Horizont, II. Beobachtungen und Theorie. *Zeitschrift für Psychologie,* **30**, 161–95.

Reimann, E. (1904). Die scheinbare Vergrösserung der Sonne und des Mondes am Horizont. *Zeitschrift für Psychologie,* **37**, 250–61.

Rentschler, I., Hilz, R., Sütterlin, C., and Noguchi, K. (1981). Illusions of filled lateral and angular extent. *Experimental Brain Research,* **44**, 154–8.

Restle, F. (1970a). Moon illusion explained on the basis of relative size. *Science,* **167**, 1092–6.

Restle, F. (1970b). Insightful amateur astronomer. *Science,* **168**, 1287.

Riccioli, G. B. (1651). *Almagestum novum, astronomiam veterem novamque complectens, etc.* Bonaniae: Haeredis V. Benatii.

Richards, W. (1977). Lessons in constancy from neurophysiology. In W. Epstein (ed.), *Stability and constancy in visual perception*, pp. 421–36. New York: Wiley.

Ridpath, I. (ed.) (1998). *Norton's star atlas and reference handbook (Epoch 2000.0)* (19th ed). Harlow: Addison Wesley Longman.

Rist, J. M. (1972). *Epicurus: an introduction.* Cambridge: Cambridge University Press.

Rivers, W. H. (1896). On the apparent size of objects. *Mind*, new series, **5**, 71–80.

Robins, B. (1761). *Mathematical tracts of the late Benjamin Robins*; published by James Wilson. London: J. Nourse.

Robinson, E. J. (1954). The influence of photometric brightness on judgments of size. *American Journal of Psychology*, **67**, 464–74.

Robinson, J. O. (1972). *The psychology of visual illusion.* London: Hutchinson University Library.

Robinson, T. M. (1987). *Heraclitus. Fragments. A text and translation with a commentary.* Toronto: University of Toronto Press.

Rock, I. (1966). *The nature of perceptual adaptation.* New York: Basic Books.

Rock, I. (1975). *An introduction to perception.* New York: Macmillan.

Rock, I. (1977). In defense of unconscious inference. In W. Epstein (ed.), *Stability and constancy in visual perception*, pp. 321–73. New York: Wiley.

Rock, I. and Ebenholtz, S. (1959). The relational determination of perceived size. *Psychological Review*, **66**, 387–401.

Rock, I. and Kaufman, L. (1962). The moon illusion, II. *Science*, **136**, 1023–31.

Rock, I. and McDermott, E. (1964). The perception of visual angle. *Acta Psychologica*, **22**, 119–14.

Roelofs, C. O. and Zeeman, W. P. C. (1957). Apparent size and apparent distance in binocular and monocular vision. *Ophthalmologica*, **133**, 188–204.

Rohault, J. (1671). *A system of natural philosophy*, 2 vols. (revised edn of 1723, ed. and trans. J. and S. Clarke). New York: Johnson Reprint Corporation, 1969.

Ronchi, V. (1957). Optics: the science of vision (trans. E. Rosen). New York: New York University Press.

Roscoe, S. N. (1977). *How big the moon, how fat the eye?* (Tech. Rep. ARL-77-2/AFOSR-77-2). Savoy: University of Illinois at Urbana-Champaign, Aviation Research Laboratory.

Roscoe, S. N. (1979). When day is done and shadows fall, we miss the airport most of all. *Human Factors*, **21**, 721–31.

Roscoe, S. N. (1982). Landing airplanes, detecting traffic, and the dark focus. *Aviation, Space, and Environmental Medicine*, **53**, 970–6.

Roscoe, S. N. (1985). Bigness is in the eye of the beholder. *Human Factors*, **27**, 615–36.

Roscoe, S. N. (1989). The zoom-lens hypothesis. In M. Hershenson (ed.), *The moon illusion*, pp. 31–57. Hillsdale, NJ: Erlbaum.

Ross, H. E. (1965). The size-constancy of underwater swimmers. *Quarterly Journal of Experimental Psychology,* **17**, 329–37.

Ross, H. E. (1967). Water, fog and the size–distance invariance hypothesis. *British Journal of Psychology,* **58**, 301–13.

Ross, H. E. (1970). Adaptation of divers to curvature distortion under water. *Ergonomics,* **13**, 489–99.

Ross, H. E. (1971). Spatial perception under water. In J. D. Woods and J. N. Lythgoe (ed.), *Underwater science,* pp. 69–101. London: Oxford University Press.

Ross, H. E. (1974). *Behaviour and perception in strange environments.* London: Allen and Unwin.

Ross, H. E. (1975). Mist, murk and visual perception. *New Scientist,* 17 June, 658–60.

Ross, H. E. (1982). *Visual-vestibular interaction in size perception.* Report to the Visual Research Trust, Department of Psychology, University of Stirling.

Ross, H. E. (1993). Do judgements of distance and greyness follow the physics of aerial perspective? In A. Garriga-Trillo, P. R. Minon, C. Garcia-Gallego, P. Lubin, J. M. Merino, and A. Vallarino (ed.), *Fechner Day '93,* pp. 233–38. Palma de Majorca, Spain: International Society for Psychophysics.

Ross, H. E. (1994a). Scaling the heights. *The Scottish Mountaineering Club Journal,* **XXXV**, No.185, 402–10.

Ross, H. E. (1994b). Active and passive head and body movements. *Behavioral and Brain Sciences,* **17**, 329–30.

Ross, H. E. (1997). On the possible relations between discriminability and apparent magnitude. *British Journal of Mathematical and Statistical Psychology,* **50**, 187–203.

Ross, H. E. (2000). Cleomedes (*c.* 1st century AD) on the celestial illusion, atmospheric enlargement and size–distance invariance. *Perception,* **29**, 853–61.

Ross, H. E. (2002). Levels of processing in the size–distance paradox. In L. R. Harris and M. Jenkin (ed.) *Levels of perception,* pp. 143–62. New York: Springer Verlag.

Ross, H. E. and Plug, C. (1998). The history of size constancy and size illusions. In V. Walsh and J. J. Kulikowski (ed.), *Perceptual constancy: why things look as they do,* pp. 499–528. Cambridge: Cambridge University Press.

Ross, H. E. and Ross, G. M. (1976). Did Ptolemy understand the moon illusion? *Perception,* **5**, 377–85.

Ross, H. E., King, S. R., and Snowden, H. (1970). Size and distance judgements in the vertical plane under water. *Psychologische Forschung,* **33**, 155–64.

Ross, J., Jenkins, B., and Johnstone, J. R. (1980). Size constancy fails below half a degree. *Nature,* **283**, 473–4.

Ruggles, C. (1999). *Astronomy in prehistoric Britain and Ireland.* New Haven: Yale University Press.

Rump, E. E. (1961). The relationship between perceived size and perceived distance. *British Journal of Psychology*, **52**, 111–24.

Sabra, A. I. (1987). Psychology versus mathematics: Ptolemy and Alhazen on the moon illusion. In E. Grant and J. E. Murdoch (ed.), *Mathematics and its applications to science and natural philosophy in the Middle Ages*, pp. 217–47. Cambridge: Cambridge University Press.

Sabra, A. I. (1989). Vol. I: *The Optics of Ibn Al-Haytham: Books I–III On direct vision*. Vol. II: *Introduction and commentary*. Studies of the Warburg Institute (ed. J. B. Trapp), Vol. 40, i. London: University of London.

Sabra, A. I. (1996). On seeing the stars, II: Ibn al Haytham's 'answers' to the 'doubts' raised by Ibn Ma'Dan. *Zeitschrift für Geschichte der Arabisch-Islamischen Wissenschaften*, Sonderdruck, Band 10. Frankfurt am Main: Institut für Geschichte der Arabisch-Islamischen Wissenschaften.

Sanford, E. C. (1898). *A course in experimental psychology. Part 1: Sensation and perception*. London: Heath.

Schaeberle, F. M. (1899). A simple physical explanation of the seeming enlargement of celestial areas near the horizon. *Astronomische Nachrichten*, **148**, 375.

Scheerer, E. (1984). Motor theories of cognitive structure: a historical review. In W. Prinz and A. F. Sanders (ed.), *Cognition and motor processes*, pp. 77–98. Berlin: Springer.

Scheerer, E. (1987). Muscle sense and innervation feelings: a chapter in the history of perception and action. In H. Heuer and A. F. Sanders (ed.), *Perspectives on perception and action*, pp. 171–94. Hillsdale, NJ: Erlbaum.

Schlesinger, F. (1913). Elliptical lunar halos. *Nature*, **91**, 110–11.

Schmidt, I. (1964). Is there a moon illusion in space? *Aerospace Medicine*, **35**, 572–5.

Schönbeck, G. L. J. (1998). *Sunbowl or symbol: models for the interpretation of Heraclitus' sun notion*. Amsterdam: Elixir Press.

Schramm, M. (1959). Zur Entwicklung der physiologischen Optik in der arabischen Litteratur. *Sudhoffs Archiv für Geschichte der Medizin und der Naturwissenschaften*, **43**, 289–316.

Schur, E. (1925). Mondtäuschung und Sehgrossenkonstanz. *Psychologische Forschung*, **7**, 44–80.

Schwartz, E. L. (1980). Computational anatomy and functional architecture of striate cortex: a spatial mapping approach to perceptual coding. *Vision Research*, **20**, 645–69.

Schwartz, R. (1994). *Vision: variations on some Berkeleian themes*. Oxford: Blackwell.

Sears, F. W. (1958). *Optics* (3rd edn). Reading, MA: Addison-Wesley.

Sedgwick, H. A. (1973). The visible horizon. *Dissertation Abstracts International 1973–74*, **34**, 1301–1302B (University Microfilms 73–22, 530).

Sedgwick, H. A. (1986). Space perception. In K. R. Boff, L. Kaufman, and J. P. Thomas (ed), *Handbook of perception and human performance*, Vol. 1: *Sensory processes and perception*, pp. 21.1–21.57. New York: Wiley.

Seneca, Lucius Annaeus (*c.* 3 BC – AD 65). *Naturales quaestiones*, Vol. I (trans. T. H. Corcoran). London: Loeb, 1971.

Shallo, J. and Rock, I. (1988). Size constancy in children: a new interpretation. *Perception*, **17**, 803–13.

Siegel, R. S. (1970). *Galen on sense perception*. Basel: Karger.

Simonelli, N. M. and Roscoe, S. N. (1979). *Apparent size and visual accommodation under day and night conditions*. Technical report Eng Psy-79-3/AFOSR-79-3. Champaigne, Illinois: Department of Psychology, University of Illinois at Urbana-Champaigne.

Simonet, P. and Campbell, M. C. W. (1990). Effect of illuminance on the directions of chromostereopsis and transverse chromatic aberration observed with natural pupils. *Ophthalmic and Physiological Optics*, **10**, 271–9.

Smith, A. M. (1982). Ptolemy's search for a law of refraction: A case-study in the classical methodology of "saving the appearances" and its limitations. *Archive for History of Exact Sciences*, **26**, 221–40.

Smith, A. M. (1996). *Ptolemy's theory of visual perception: an English translation of the Optics with introduction and commentary*. Transactions of the American Philosophical Society, **86**, Part 2. Philadelphia: American Philosophical Society.

Smith, N. (1905). Malebranche's theory of the perception of distance and magnitude. *British Journal of Psychology*, **1**, 191–204.

Smith, O. W., Smith, P. C., Geist, C. C., and Zimmerman, R. R. (1978). Apparent size contrasts of retinal images and size constancy as determinants of the moon illusion. *Perceptual and Motor Skills*, **46**, 803–8.

Smith, R. (1985). Vergence eye movement responses to whole-body linear acceleration stimuli in man. *Ophthalmological and Physiological Optics*, **5**, 303–11.

Smith, Robert. (1738). *A compleat system of opticks*. Cambridge.

Smith, W. M. (1952). Gilinsky's theory of visual size and distance. *Psychological Review*, **59**, 239–43.

Solhkhah, N. and Orbach, J. (1969). Determinants of the magnitude of the moon illusion. *Perceptual and Motor Skills*, **29**, 87–98.

Sterneck, R. von. (1907). *Der Sehraum auf Grund der Erfahrung*. Leipzig: J. A. Barth.

Sterneck, R. von. (1908). Die referenzflächentheorie der scheinbaren Grösse der Gestirne. *Zeitschrift für Psychologie*, **46**, 1–22.

Storch, E. (1903). Der Wille und das räumliche moment in Wahrnehmung und Vorstellung. *(Pflügers) Archiv für Physiologie*, **95**, 305–45.

Strabo (*c.* 63 BC – AD 25). *The geography* (trans. H. D. P. Lee). Loeb Classical Library, London: Heinemann, 1962.

Stroobant, P. (1884). Sur l'agrandissement apparent des constellations, du soleil et de la lune à l'horizon. *Bulletin de l'Académie Royale de Belgique*, 3rd series, **8**, 719–31.

Stroobant, P. (1928). Sur l'agrandissement apparent des constellations, du soleil et de la lune à l'horizon (troisième note). *Bulletin de l'Académie Royale de Belgique*, Classe des Sciences, 5th series, **14**, 91–108.

Strous, Dr (1999). *Special full moon.* http://louis.lmsal.com/PR/specialmoon.html. Downloaded 15 February 2001.

Sundet, J. M. (1972). The effect of pupil size variations on the color stereoscopic phenomenon. *Vision Research*, **12**, 1027–32.

Sundet, J. M. (1978). Effects of colour on perceived depth. Review of experiments and evaluation of theories. *Scandinavian Journal of Psychology*, **19**, 133–43.

Suzuki, K. (1991). Moon illusion simulated in complete darkness: planetarium experiment reexamined. *Perception and Psychophysics*, **49**, 349–54.

Suzuki, K. (1998). The role of binocular viewing in a spacing illusion arising in a darkened surround. *Perception*, **27**, 355–61.

Szily, A. von (1905). Bewegungsnachbild und Bewegungskontrast. *Zeitschrift für Psychologie und Physiologie der Sinnesorgane*, **38**, 81–154.

Talbot, P. A. (1969). *The peoples of southern Nigeria*, 4 vols. (new impression). London: Frank Cass (originally published 1926).

Taylor, D. W. and Boring, E. G. (1942). The moon illusion as a function of binocular regard. *American Journal of Psychology*, **55**, 189–201.

Taylor, F. V. (1941). Change in size of the after-image induced in total darkness. *Journal of Experimental Psychology*, **29**, 75–80.

Taylor, I. L. and Sumner, F. C. (1945). Actual brightness and distance of individual colors when their apparent distance is held constant. *Journal of Psychology*, **19**, 79–85.

Teghtsoonian, M. (1965). The judgment of size. *American Journal of Psychology*, **78**, 392–402.

Teghtsoonian, M. and Beckwith, J. B. (1976). Children's size judgments when size and distance vary: is there a developmental trend to overconstancy? *Journal of Experimental Child Psychology*, **22**, 23–39.

Teghtsoonian, M. and Teghtsoonian R. (1970). Scaling apparent distance in a natural outdoor setting. *Psychonomic Science*, **21**, 215–16.

Thompson, R. C. (1900). *The reports of the magicians and astrologers of Nineveh and Babylon.* London: Luzac and Co.

Thor, D. H., Winters, J. J., and Hoats, D. L. (1969). Vertical eye movement and space perception: a developmental study. *Journal of Experimental Psychology*, **82**, 163–7.

Thor, D. H., Winters, J. J., and Hoats, D. L. (1970). Eye elevation and visual space in monocular regard. *Journal of Experimental Psychology*, **86**, 246–9.

Thorndike, L. (1923–58). *A history of magic and experimental science*, 8 vols. New York: Columbia University Press.

Thouless, R. H. (1968). Apparent size and distance in vision through a magnifying system. *British Journal of Psychology*, **59**, 111–18.

Titchener, E. B. (1901). *Experimental psychology: a manual of laboratory practice*, Vol. 1: *Qualitative experiments; Part 1. Student's manual.* London: Macmillan.

Todd, R. (ed.) (1990). *Cleomedis caelestia (meteora).* Leipzig: Teubner.

Tolansky, S. (1964). *Optical illusions.* London: Pergamon Press.

Tomlinson, C. (*c.* 1862). *The dew-drop and the mist.* London: Society for Promoting Christian Knowledge.

Toomer, G. J. (1974). Hipparchus on the distances of the sun and moon. *Archive for History of Exact Sciences*, **14**, 126–42.

Trehub, A. (1991). *The cognitive brain.* Cambridge, MA: MIT Press.

Tronick, E. and Hershenson, M. (1979). Size–distance perception in pre-school children. *Journal of Experimental Child Psychology*, **27**, 166–84.

Trotter, A. P. (1938). The apparent size of the sun. *Nature*, **141**, 123.

Tscherning, M. (1904). *Physiologic optics* (2nd edn; trans. C. Weiland). Philadelphia: The Keystone (The Organ of the Jewelry and Optical Trades).

Turnbull, C. M. (1961). Some observations regarding the experiences and behavior of the BaMbuti Pygmies. *American Journal of Psychology*, **74**, 304–8.

Van de Geer, J. P. and Zwaan, E. J. (1964). Size constancy as dependent upon angle of regard and spatial direction of stimulus-object. *American Journal of Psychology*, **77**, 563–75.

Van Eyl, F. P. (1968). Vestibular hypothesis for the moon illusion. In *Proceedings of the 76th Annual Convention of the American Psychological Association*, **3**, 87–8. Washington, D.C.: APA.

Van Eyl, F. P. (1972). Induced vestibular stimulation and the moon illusion. *Journal of Experimental Psychology*, **94**, 326–8.

Vitruvius Pollio (1st century BC). *De architectura* (trans. F. Granger). New York: G. P. Putnam's Sons, 1931.

Vitruvius Pollio (1st century BC). *On architecture*, 2 vols. (ed. and trans. F. Granger). Cambridge, MA: Harvard University Press, 1970.

Wade, N. J. (1998a). Light and sight since antiquity. *Perception*, **27**, 637–70.

Wade, N. J. (1998b). *A natural history of vision.* Cambridge, MA: MIT Press.

Wade, N. J. and Day, R. H. (1967). Tilt and centrifugation in changing the direction of body-force. *American Journal of Psychology*, **80**, 637–9.

Wagner, M., Baird, J. C., and Fuld, K. (1989). Transformation model of the moon illusion. In M. Hershenson (ed.), *The moon illusion*, pp. 147–65. Hillsdale, NJ: Erlbaum.

Walker, B. H. (1978a). The moon illusion: a review – part 1. *Optical Spectra*, **12** (1), 68–9.

Walker, B. H. (1978b). The moon illusion: a review – part 2. *Optical Spectra*, **12** (2), 64–5.

Walker, E. (1804). On the apparent size of the horizontal moon. *Journal of Natural Philosophy, Chemistry and the Arts*, **9**, 164–6.

Walker, E. (1805). On the apparent size of the horizontal moon. *Journal of Natural Philosophy, Chemistry and the Arts*, **10**, 105–11.

Walker, E. (1806). Abridgment of certain papers written on the apparent magnitude of the horizontal moon. *Philosophical Magazine*, **24**, 240–4.

Walker, E. (1807). On the phenomena of the horizontal moon. *Philosophical Magazine*, **29**, 65–8.

Walker, J. (1997). *Inconstant moon: the moon at perigee and apogee.* http://www.fourmilab.ch/earthview/moon_ap_per.html. Downloaded 15 February 2001.

Wallach, H. (1948). Brightness constancy and the nature of achromatic colors. *Journal of Experimental Psychology*, **38**, 310–24.

Wallach, H. (1962). On the moon illusion (letters to the editor). *Science*, **137**, 900–2.

Wallach, H. and Berson, E. (1989). Measurements of the illusion. In M. Hershenson (ed.), *The moon illusion*, pp. 287–97. Hillsdale, NJ: Erlbaum.

Wallach, H. and McKenna, V. V. (1960). On size-perception in the absence of cues for distance. *American Journal of Psychology*, **73**, 458–60.

Wallis, J. (1687). [Comments on Molyneux]. *Philosophical Transactions of the Royal Society of London*, **16**, 323–9.

Wallis, W. A. (1935). The influence of color on apparent size. *Journal of General Psychology*, **13**, 193–9.

Walsh, V. and Kulikowski, J. J. (ed.) (1998). *Perceptual constancy: why things look as they do.* Cambridge: Cambridge University Press.

Warner, B. (1996). Traditional astronomical knowledge in Africa. In C. Walker (ed.), *Astronomy before the telescope*, pp. 304–17. London: British Museum Press.

Warren, R. and Owen, D. H. (1980). The horizon, ecological optics, and the moon illusion. *Bulletin of the Psychonomic Society*, **10**, 167.

Warren, W. H. and Whang, S. (1987). Visual guidance of walking through apertures: body-scaled information for affordances. *Journal of Experimental Psychology: Human Perception and Performance*, **13**, 371–83.

Weale, R. A. (1975). Apparent size and contrast. *Vision Research*, **15**, 949–55.

Weber, E. H. (1834). *De pulsu, resorptione, auditu et tactu. Annotationes anatomicae et physiologicae.* Leipzig: Koehler. Trans. in H. E. Ross and D. J. Murray (1996). *E. H. Weber on the tactile senses* (2nd edn). Hove: Erlbaum (UK)/Taylor and Francis.

Weber, E. H. (1846). *Der Tastsinn und das Gemeingefühl.* Trans. in H. E. Ross and D. J. Murray (1996). *E. H. Weber on the tactile senses* (2nd edn). Hove: Erlbaum (UK)/Taylor and Francis.

Welch, R. B. (1978). *Perceptual modification: adapting to altered sensory environments.* New York: Academic Press.

Welch, R. B. (1986). Adaptation of space perception. In K. R. Boff, L. Kaufman, and J. P. Thomas (ed.), *Handbook of perception and human performance*, Vol. I: *Sensory processes and perception*, Chapter 24, pp. 1–45. New York: Wiley.

Wenning, C. J. (1985). New thoughts on understanding the moon illusion. *Planetarian*, **14** (December), No. 4. Retrieved 28 January 1999 from the World Wide Web: http://www.GriffithObs.org/IPSMoonIllus.html

Wheatstone, C. (1852). Contributions to the physiology of vision – part the second. On some remarkable, and hitherto unobserved, phenomena of binocular vision. *Philosophical Transactions of the Royal Society*, **142**, 1–17.

Whiteside, T. C. D. and Campbell, F. W. (1959). Size constancy effect during angular and radial acceleration. *Quarterly Journal of Experimental Psychology*, **11**, 249.

Wiedemann, E. (1914). Fragen aus dem Gebiet der Naturwissenschaften, gestellt von Friedrich II. *Archiv für Kulturgeschichte*, **11**, 483–5.

Wilkins, J. (1638). *The discovery of a world in the moone*. London: Sparke and Forrest. Scholars' Facsimiles and Reprints, New York: Delmar, 1973.

Willats, J. (1997). *Art and representation: new principles in the analysis of pictures*. Princeton, NJ: Princeton University Press.

Witkin, H. A. (1964). Uses of the centrifuge in studies of the orientation of space. *American Journal of Psychology*, **77**, 499–502.

Witte, H. (1918). Ueber den Sehraum. Vorläufige Mitteilung über eine gemeinsam mit Herrn E. Laqueur unternommene Arbeit. *Physikalische Zeitschrift*, **19**, 142–51.

Witte, H. (1919a). Ueber den Sehraum. Zweite Mitteilung: Zur Frage nach der scheinbaren Vergrösserung des Mondes usw. am Horizont. *Physikalische Zeitschrift*, **20**, 61–4.

Witte, H. (1919b). Ueber den Sehraum. Dritte Mitteilung: Zur scheinbaren Grösse des Mondes. *Physikalische Zeitschrift*, **20**, 114–20.

Witte, H. (1919c). Ueber den Sehraum. Vierte Mitteilung: Scheinbare Grösse und scheinbare Vergrösserung des Mondes. *Physikalische Zeitschrift*, **20**, 126–7.

Wolbarsht, M. L. and Lockhead, G. R. (1985). Moon illusion: a new perspective. *Applied Optics* (USA), **24**, 1844–7.

Wood, R. J., Zinkus, P. W., and Mountjoy, P. T. (1968). The vestibular hypothesis of the moon illusion. *Psychonomic Science*, **11**, 356.

Woodworth, R. S. and Schlosberg, H. (1954). *Experimental psychology*. New York: Holt.

Wundt, W. (1873). *Grundzüge der physiologischen Psychologie*. Leipzig: W. Engelmann (6th edn 1908–11).

Ye, M., Bradley, A., Thibos, L. N., and Zhang, X. (1991). Interocular differences in transverse chromatic aberration determine chromostereopsis for small pupils. *Vision Research*, **31**, 1787–96.

Young, T. (1807a). *A course of lectures on natural philosophy and the mechanical arts* (new edn). London: Taylor and Walton, 1845.

Young, T. (1807b). *A course of lectures on natural philosophy and the mechanical arts*, Vol.1: Lecture XXXVIII, On vision. London: Johnson. Reprinted in *The sources of science*, No. 82. New York: Johnson Reprint Corporation, 1971.

Zehender, W. von (1899). Die Form des Himmelsgewölbes und das Grösser-Erscheinen der Gestirne am Horizont. *Zeitschrift für Psychologie*, **20**, 353–7.

Zehender, W. von (1900). Die Form des Himmelsgewölbes und das Grösser-Erscheinen der Gestirne am Horizont. *Zeitschrift für Psychologie*, **24**, 218–84.

Zeilik, M. and Gregory, S. A. (1998). *Introductory astronomy and astrophysics* (4th edn). Fort Worth: Saunders College Publishing.

Zeno, T. (1862). On the changes in the apparent size of the moon. *Philosophical Magazine*, **24**, 390–2.

Zinkus, P. W. and Mountjoy, P. T. (1969). The effect of head position on size discrimination. *Psychonomic Science*, **14**, 80.

Zinkus, P. W. and Mountjoy, P. T. (1978). The effects of information and training on the discrimination of size in different planes of space. *Psychological Record*, **28**, 383–90.

Zoth, O. (1899). Ueber den Einfluss der Blickrichtung auf die scheinbare Grösse der Gestirne und die scheinbare Form des Himmelsgewölbes. *(Pflügers) Archiv für Physiologie*, **78**, 363–401.

Zoth, O. (1902). Bemerkungen zu einer alten 'Erklärung' und zwei neuen Arbeiten, Betreffend die scheinbare Grösse der Gestirne und Form des Himmelsgewölbes. *(Pflügers) Archiv für Physiologie*, **88**, 201–24.

Name Index

Subject Index